MIRACLE CURE

MIRACLE CURE

The Creation of Antibiotics and the
Birth of Modern Medicine

WILLIAM ROSEN

VIKING

VIKING
An imprint of Penguin Random House LLC
375 Hudson Street
New York, New York 10014
penguin.com

LIBRARY OF CONGRESS CATALOGING-IN-PUBLICATION DATA
Names: Rosen, William, 1955– author.
Title: Miracle cure : the creation of antibiotics and the birth of modern
medicine / William Rosen.
Description: New York, New York : Viking, [2017] | Includes bibliographical
references and index.
Identifiers: LCCN 2016029488 (print) | LCCN 2016030953 (ebook) | ISBN
9780525428107 (hardcover) | ISBN 9780698184107 (ebook)
Subjects: | MESH: Anti-Bacterial Agents—history | History, 19th Century |
History, 20th Century | History, 21st Century
Classification: LCC RM409 (print) | LCC RM409 (ebook) | NLM QV 11.1 | DDC
615.7/922—dc23
LC record available at https://lccn.loc.gov/2016029488

Printed in the United States of America
1   3   5   7   9   10   8   6   4   2

Set in Aldus LT Std
Designed by Alissa Rose Theodor

CONTENTS

MIRACLE CURE

PROLOGUE

"Five and a Half Grams"

The Smithsonian Institution's National Museum of American History is home to more than ten million items, from Julia Child's sauté pan to a twenty-foot-wide concrete chunk of the original Route 66. Less of a draw for crowds, but far more historically significant, is a thirty-page-long document—a hospital chart—covering a month in the life of a single patient: Anne Miller, whose fourth pregnancy ended in a miscarriage on Valentine's Day, 1942. The written record of her hospital stay begins almost immediately thereafter, when, after experiencing severe chills and temperatures spiking to more than 106°, she was transferred from the maternity wing of New Haven Hospital to the intensive care unit. She had contracted *hemolytic streptococcal septicemia*; informally, blood poisoning.

Her condition, already poor, deteriorated. Each day, a box on Mrs. Miller's chart recorded the number of bacterial colonies found in her blood. The one for March 1 contains the symbol ∞: infinity. She was given blood transfusions, rattlesnake serum, and even the new sulfa drugs. Nothing worked.

On March 11, her doctor, John Bumstead, approached another of his patients, John Fulton, a neurophysiologist from Yale University, who had been hospitalized for a bronchial infection he acquired on a survey of California laboratories. Dr. Bumstead knew that Fulton was a close friend of Howard Florey, an Australian pathologist working in London; so close, in fact, that Fulton had agreed to care for the Florey children for the duration of the war. Bumstead also knew that Dr. Florey was the world's leading authority on, and advocate for, the therapeutic potential of a compound first discovered fourteen years before: penicillin.

1

Fulton agreed to call Florey's colleague, Norman Heatley, who worked at the Merck & Co. pharmaceutical plant in Rahway, New Jersey. Heatley, in turn, importuned Randolph Major, head of Merck's research labs, to secure a quantity of penicillin, then one of the rarest substances in the world, and one with such a high priority for the war effort that its release required approval from the National Research Council in Washington, DC. One of the NRC's senior members was Major's boss, George W. Merck.

The calls worked. Early on the morning of Saturday, March 14, a glass vial containing five and a half grams of brown powder—half the amount of penicillin then available in the entire country—left Rahway. By noon, it was delivered to Dr. Bumstead, who, with no way of knowing the size of a recommended dosage, invented one on the spot. He dissolved a portion of the powder in saline solution and gave Mrs. Miller a test dose intravenously at 3:30 P.M. Once she seemed to tolerate it, he gave her another dose every four hours through the night.

By Sunday, Mrs. Miller's temperature had dropped to 99°. The fifth person in the world to receive a treatment of penicillin, and the first whose life was saved by it, was sitting up and eating. She would continue doing so for the next fifty-seven years. The eons-long war between humanity and infectious disease had a new order of battle, one that organized research hospitals, university laboratories, pharmaceutical corporations, and national governments into an army. The antibiotic age had arrived.

Some revolutions are only visible from a comfortable distance in time. Not so antibiotics. Through the first four decades of the twentieth century, and well into the fifth, the second leading cause of death in the United States was pneumonia, overwhelmingly caused by *Streptococcus pneumoniae*, a first cousin to the bacterium that nearly took Anne Miller's life. The sixth deadliest killer, nearly every year, was *Mycobacterium tuberculosis*. By 1955, pneumonia had fallen to sixth place. Tuberculosis wasn't even on the list.

To anyone born in the last sixty years—and certainly anyone who has visited a hospital during those years—the conquest of so many mortal diseases is commonplace. To people who lived through the upheaval, it was astonishing. The physician and writer Lewis Thomas, writing forty years after his own medical education in the late 1930s, recalled that a doctor, in those days, could set a bone, deliver a baby, and predict the course of an illness. Actually treating someone, however, was almost entirely palliative: making the patient as comfortable as possible—when, that is, they could even do that. There are still millions of people living today who can remember when cutting a finger on a piece of barbed wire could mean—often *did* mean—an excruciating death from tetanus, bacteremia, or sepsis. Until the first antibiotics, medicine remained the oldest art. It had yet to become, in Thomas's words, the "youngest science."

The twentieth century is rightly acknowledged as an era of unimaginably rapid scientific advances, from internal combustion engines to telephones to digital computers. Famously, less than forty years separated Einstein's relativity equations from Hiroshima; James Chadwick only discovered the neutron in 1932. Antibiotics moved from experiment to application even faster; all the great families of antibacterial therapies—the sulfa drugs, beta-lactams (like penicillin), chloramphenicol, tetracycline, erythromycin, streptomycin, and the cephalosporins—appeared in a span of less than ten years.

A hundred characters figured in the birth of antibiotics: physicians who weren't scientists, scientists who weren't physicians, government bureaucrats, philanthropists, and industrialists both venal and visionary. Their story played out in university laboratories, agricultural research stations, battlefield hospitals, and the boardrooms of huge multinational corporations. The origin story of antibiotics, and of modern medicine itself, was centuries in gestation.

The war between *Homo sapiens* and infectious disease, after all, was under way for untold millennia before humanity was able to fight back.

ONE

"All the Worse for the Fishes"

O n the morning of Saturday, December 14, 1799, George Washington awakened before dawn, and told his wife, Martha, he was so sick that he could barely breathe.

The indicated treatment was straightforward enough. By the time the sun had risen, Washington's overseer, George Rawlins, "who was used to bleeding the people," had opened a vein in Washington's arm from which he drained approximately twelve ounces of his employer's blood. Over the course of the next ten hours, two other doctors—Dr. James Craik and Dr. Elisha Dick—bled Washington four more times, extracting as much as one hundred additional ounces.

Removing at least 60 percent of their patient's total blood supply was only one of the curative tactics used by Washington's doctors. The former president's neck was coated with a paste composed of wax and beef fat mixed with an irritant made from the secretions of dried beetles, one powerful enough to raise blisters, which were then opened and drained, apparently in the belief that it would remove the disease-causing poisons. He gargled a mixture of molasses, vinegar, and butter; his legs and feet were covered with a poultice made from wheat bran; he was given an enema; and, just to be on the safe side, his doctors gave Washington a dose of calomel—mercurous chloride— as a purgative.

Unsurprisingly, none of these therapeutic efforts worked. By 10:00 P.M., America's first president knew he was dying. His last words were, "I am just going! Have me decently buried, and do not let my body be put into the vault less than three days after I am dead. Do you understand me? 'Tis well!"*

Less than twenty-two years later, another world-historic figure had his final encounter with early nineteenth-century medicine. Napoleon Bonaparte, exiled to Longwood House on the South Atlantic island of St. Helena after his defeat at Waterloo in 1815, experienced bouts of abdominal pain and vomiting for months, while four different physicians (each of whom wrote a memoir about their famous patient) treated him by administering hundreds of enemas and regularly dosing him with the powerful emetic known chemically as antimony potassium tartrate—not, perhaps, the best treatment for a patient already weak from vomiting. The onetime emperor of France breathed his last on May 5, 1821.

Historians with a morbid bent have produced thousands of pages of speculation on the diseases that killed two of the most famous men who ever lived. Today, the prevailing retrospective diagnosis for Washington is that he was dispatched by an infection of the epiglottis, probably caused by the tiny organism known as *Haemophilus influenzae* type b, the pathogen that also causes bacterial meningitis. A popular minority opinion is that Washington died from PTA, or peritonsillar abscess, a strep infection that creates an abscess under the tonsil that swells with pus until it actually strangles the patient. (The other name for PTA is "quinsy" or "quinsey," from the Greek word that means "to strangle a dog.") One thing that *didn't* kill Washington, despite its mention in just about every biography of the man, was his tour of Mount Vernon in cold, wet weather during the days leading up to his death, and his decision to dine with friends wearing still-wet clothing on the night of December 13. Infectious diseases aren't caused by catching a chill.

The debate about the cause of Napoleon Bonaparte's death is, likewise, fueled by enough raw material that it seems likely to go on

---

* Like thousands of others in the late eighteenth and early nineteenth centuries, the "Father of His Country" faced death with equanimity, but not the prospect of being buried alive. Hundreds of designs for so-called "safety coffins," fitted with bells or other signaling devices, were patented from 1780 on.

forever. The initial autopsy concluded that *l'empereur* had died of stomach cancer, the same disease that killed Napoleon's father in 1785. Hepatitis has its advocates, as does the parasitic disease known as schistosomiasis, which Napoleon is thought to have acquired during his Egyptian campaign of 1798. Neither is as popular among amateur historians as arsenic poisoning, either as a murder weapon or an accident caused by exposure to wallpaper more or less saturated in the stuff.

An equally honest answer for both former generals is that they died of *iatrogenesis*. Bad medicine. Or, more accurately, *heroic* medicine.

The era of heroic medicine is generally used to describe the period, roughly 1780–1850, during which medical education and practice was highly interventional, even when the interventions did at least as much harm as good. The dates are slightly deceptive. Medical practice, from Hippocrates to Obamacare, has constantly oscillated between interventional and conservative approaches; the perfect point of balance is a moving target for physicians, and seems likely always to be.

Consider the persistence of the humoral theory of disease. Bloodletting, for example, was first popularized by the second-century Greek physician Galen of Pergamon as a way of balancing the four humors: blood, phlegm, and black and yellow bile. The doctrine that gave a vocabulary for distinct individual temperaments based on the relative amounts of these bodily fluids—sanguine personalities had a high level of blood; bile led to biliousness—was originally a guide to medical practice: Too much bile caused fevers, while too much phlegm resulted in epilepsy.

Humoral doctrine, in one form or another, dominated Western medicine for nearly two thousand years. It didn't persist because following its dictates improved the chances that patients would recover from disease, nor even because it was an accurate guide to physiology. Blood, in classic humoralism, was produced by the liver; what the humoral physician believed to be "black bile" was likely blood that had been exposed to oxygen. As much as anything else, the appeal of humoralism seems to have been the belief that, since health was a sign of balance, disease must represent an imbalance of *something*. It also reinforced a nearly universal belief in elementalism, which suggested that all phenomena could be reduced to interactions among fundamental elements like air, earth, water, and fire.

The real secret to humoralism's durability was the lack of a superior alternative. And it was nothing if not durable. Humoral balancing was still being recommended in the 1923 edition of *The Principles and Practices of Medicine* by Sir William Osler, one of the four founders of the Johns Hopkins School of Medicine. The sixth-century Byzantine physician Alexander of Tralles may have treated his patients with powerful alkaloid extracts like atropine and belladonna, purged them with verdigris (copper[II] acetate), and sedated them with opium. But he also cared for them in hospitals whose primary function was not treatment, but support: making patients as comfortable as possible while waiting for either recovery or death. The best-known words of the fifth-century B.C. Father of Medicine, Hippocrates, are "do no harm" and "nature is the best healer."*

If a further reminder is needed that the practice of heroic medicine began long before the eighteenth century, its greatest icon lived nearly two centuries before his successors turned him into a cult figure. The Swiss German physician, astrologer, and master of all things occult named Philippus Aureolus Theophrastus Bombastus von Hohenheim—as a kindness to modern readers, he is generally remembered by the honorific Paracelsus—was, like Galen, a protoscientist: a careful observer of nature limited by the lack of a mechanism for testing his hypotheses experimentally. Though he recognized the inadequacies of the humoral theories of Galen, he substituted an equally unlikely schema built on the balance of three different elements: mercury, sulfur, and salt. From the sixteenth century on, mercury especially became a remarkably popular remedy for virtually

---

* Neither phrase appears in that precise language in either the Hippocratic Oath or other works of Hippocrates—the original oath is usually translated as ". . . never do harm to anyone"—but there's little doubt that Hippocrates subscribed to the sentiments they describe.

every medical condition. In 1530, Paracelsus recommended mercury with such enthusiasm that he inspired the Austrian physician Gerard van Swieten to prescribe it in the form of mercuric oxide—more soluble in water and, therefore, even more toxic than the mercurous chloride known as calomel prescribed by Washington's physicians—as a cure for syphilis. One syphilis treatment from 1720 called for four doses of calomel over three days, separated by a modest amount of bleeding—only a pint or so. Concoctions that contained mercury remained part of the materia medica for centuries because of a confusion between potency and effectiveness. Doctors cheered when patients exhibited the ulcerated gums and uncontrollable salivation that are the classic signs of mercury poisoning, as evidence that the medicine was clearly working.

Mercury therapy was only one of a collection of techniques and beliefs that seem, in retrospect, ghoulish in the extreme. True, some areas of medical knowledge had increased dramatically in the sixteen centuries after Galen. The Brussels-born physician Andreas Vesalius, the first European physician allowed to dissect human corpses, revolutionized anatomy; William Harvey discovered that blood circulated from the heart to the extremities and back again. Even the enthusiasm for mercury, and a number of other toxic substances, wasn't complete nonsense. As a more scientific group of physicians would soon demonstrate, mercury actually *does* kill some very nasty disease-causing pathogens. The great weakness of Washington's and Napoleon's physicians wasn't ineptitude—they were probably the most skilled men on the planet when it came to making their patients bleed or vomit—but theory. Eighteenth-century physicians knew as little about the causes of disease as a cat knows about calculus, and certainly no more than their predecessors had known in the second century. A doctor could set a broken bone, perform elaborate-though-useless tests, and comfort the dying, but hardly anything else. As the eighteenth century turned into the nineteenth, the search for reliable and useful treatments for disease had been under way for millennia, with no end in sight.

Benjamin Rush, the most famous physician of the newly independent United States, dosed hundreds of patients suffering from a yellow fever outbreak in 1793 Philadelphia with mercury. Rush also treated patients exhibiting signs of mental illness by blistering, the same procedure used on Washington in extremis. One 1827 recipe for a "blistering plaster" should suffice:

> Take a purified yellow Wax, mutton Suet, of each a pound; yellow Resin, four ounces; Blistering flies in fine powder, a pound. [The active ingredient of powdered "flies" is cantharidin, the highly toxic irritant secreted by many beetles, and M. cantharides, Spanish fly.] Melt the wax, the suet, and the resin together, and a little before they concrete in becoming cold, sprinkle in the blistering flies and form the whole into a plaster. . . . Blistering plasters require to remain applied [typically to the patient's neck, shoulder, or foot] for twelve hours to raise a perfect blister; they are then to be removed, the vesicle is to be cut at the most depending part . . .

Benjamin Rush was especially fond of using such plasters on his patient's shaven skull so that "permanent discharge from the neighborhood of the brain" could occur. He also developed the therapy known as "swinging," strapping his patients to chairs suspended from the ceiling and rotated for hours at a time. Not for Rush the belief that nature was the most powerful healer of all; he taught his medical students at the University of Pennsylvania to "always treat nature in a sick room as you would a noisy dog or cat."

When physicians' only diagnostic tools were eyes, hands, tongue, and nose, it's scarcely a surprise that they attended carefully to observable phenomena like urination, defecation, and blistering. As late as 1862, Dr. J. D. Spooner could write, "Every physician of experience can recall cases of internal affections [sic] which, after the administration of a great variety of medicines, have been unexpectedly relieved by an eruption on the skin." To the degree therapeutic substances were classified at all, it wasn't by the diseases they treated, but by their most obvious functions: promoting emesis, narcosis, or diuresis.

Heroic medicine was very much a creature of an age that experienced astonishing progress in virtually every scientific, political, and technological realm. The first working steam engine had kicked off the Industrial Revolution in the first decades of the eighteenth century. Between 1750 and 1820, Benjamin Franklin put electricity to work for the first time, Antoine Lavoisier and Joseph Priestley discovered oxygen, Alessandro Volta invented the battery, and James Watt the separate condenser. Thousands of miles of railroad track were laid to carry steam locomotives. Nature was no longer a state to be humbly accepted, but an enemy to be vanquished; physicians, never very humble in the first place, were easily persuaded that all this newfound chemical and mechanical knowledge was an arsenal for the conquest of disease.

And heroic efforts "worked." That is, they reliably did *something*, even if the something was as decidedly unpleasant as vomiting or diarrhea. Whether in second-century Greece or eighteenth-century Virginia (or, for that matter, twenty-first-century Los Angeles), patients expect action from their doctors, and heroic efforts often succeeded. Most of the time, patients got better.

It's hard to overstate the importance of this simple fact. Most people who contract any sort of disease improve because of a fundamental characteristic of Darwinian natural selection: The microorganisms responsible for much illness and virtually all infectious disease derive no long-term evolutionary advantage from killing their hosts. Given enough time, disease-causing pathogens almost always achieve a modus vivendi with their hosts: sickening them without killing them.* Thus, whether a doctor gives a patient a violent emetic or a cold compress, the stomachache that prompted the intervention is likely to disappear in time.

Doctors weren't alone in benefiting from the people-get-better phenomenon, or, as it is known formally, "self-limited disease." Throughout

---

* In fact, the level of virulence of any disease is a fairly good proxy for the time during which humans have been exposed to it; members of societies that have lived with *Varicella zoster*—the virus that causes chicken pox—for millennia get sick for a few weeks, both because the host's defenses grow more skilled with time and because the pathogens grow less dangerous, while populations encountering it for the first time die in large numbers.

the eighteenth and early nineteenth centuries, practitioners of what we now call alternative medicine sprouted like mushrooms all over Europe and the Americas: Herbalists, phrenologists, hydropaths, and homeopaths could all promise to cure disease at least as well as regular physicians. The German physician Franz Mesmer promoted his theory of animal magnetism, which maintained that all disease was due to a blockage in the free flow of magnetic energy, so successfully that dozens of European aristocrats sought his healing therapies.*

The United States, in particular, was a medical free market gone mad; by the 1830s, virtually no license was required to practice medicine anywhere in the country. Most practicing physicians were self-educated and self-certified. Few ever attended a specialized school or even served as apprentices to other doctors. Prescriptions, as then understood, weren't specific therapies intended for a particular patient, but recipes that druggists compounded for self-administration by the sick. Pharmacists frequently posted signs advertising that they supplied some well-known local doctor's formulations for treating everything from neuralgia to cancer. Doctors didn't require a license to sell or administer drugs, except for so-called ethical drugs—the term was coined in the middle of the nineteenth century to describe medications whose ingredients were clearly labeled—which were compounds subject to patent and assumed to be used only for the labeled purposes. Everything else, including proprietary and patent medicines (just to confuse matters, like "public schools" that aren't public, "patent" medicines weren't patented), was completely unregulated, a free-for-all libertarian dream that supplemented the Hippocratic Oath with *caveat patiens*: "Let the patient beware."†

The historical record isn't reliable when it comes to classifying causes of death, even in societies that were otherwise diligent about record-

---

* His practice was eventually debunked by an all-star commission appointed by King Louis XVI that included Antoine Lavoisier and Benjamin Franklin.
† It is no coincidence that the Latin root for "patient" is *patior*: "I am suffering."

ing dates, names, and numbers of corpses. As a case in point, the so-called Plague of Athens that afflicted the Greek city in the fifth century B.C.E. was documented by Thucydides himself, but no one really knows what caused it, and persuasive arguments for everything from a staph infection to Rift Valley fever are easily found. That such a terrifying and historically important disease outbreak remains mysterious to the most sophisticated medical technology of the twenty-first century underlines the problem faced by physicians—to say nothing of their patients—for millennia. Only a very few diseases even had a well-understood path of transmission. From the time the disease first appeared in Egypt more than 3,500 years ago, no one could fail to notice that smallpox scabs were themselves contagious, and contact with them was dangerous.* Similarly, the vector for venereal diseases like gonorrhea and syphilis—which probably originated in a nonvenereal form known as "bejel"—isn't a particularly daunting puzzle: Symptoms appear where the transmission took place. Those bitten by a rabid dog could have no doubt of what was causing their very rapid death.

On the other hand, the routes of transmission for some of the most deadly diseases, including tuberculosis, cholera, plague, typhoid fever, and pneumonia, were utterly baffling to their sufferers. Bubonic plague, which killed tens of millions of Europeans in two great pandemics, one beginning in the sixth century A.D., the other in the fourteenth, is carried by the bites of fleas carried by rats, but no one made the connection until the end of the nineteenth century. The Italian physician Girolamo Fracastoro (who not only named syphilis, in a poem entitled "Syphilis sive morbus Gallicus," but, in an excess of anti-Gallicism, first called it the "French disease") postulated, in 1546, that contagion was a "corruption which . . . passes from one thing to another and is originally caused by infection of imperceptible particles" that he called *seminaria*: the seeds of contagion. Less

---

* Also, that exposure conferred lifetime immunity, which is why vaccination against smallpox was being practiced in Asia centuries before Edward Jenner's 1796 inoculation of an eight-year-old boy named James Phipps (with lesions containing cowpox— *vaccinia*—rather than the *variola* virus that caused smallpox).

presciently, he also argued that the particles only did their mischief when the astrological signs were in the appropriate conjunction, and preserved Galen's humoral theory by suggesting that different seeds have affinities for different humors: the seed of syphilis with phlegm, for example. As a result, Fracastoro's "cures" still required expelling the seeds via purging and bloodletting, and his treatments were very much part of a medical tradition thirteen centuries old.

However, while seventeenth-century physicians (and "natural philosophers") failed to find a working theory of disease, they were no slouches when it came to collecting data about the subject. The empiricists of the Age of Reason were all over the map when it came to ideas about politics and religion, but they shared an obsessive devotion to experiment and observation. Their worldview, in practice, demanded the rigorous collection of facts and experiences, well in advance of a theory that might, in due course, explain them.

In the middle of the seventeenth century, the English physician Thomas Sydenham attempted a taxonomy of different diseases afflicting London. The haberdasher turned demographer John Graunt detailed the number and—so far as they were known—the causes of every recorded death in London in 1662, constructing the world's first mortality tables.* The French physician Pierre Louis examined the efficacy of bloodletting on different populations of patients, thus introducing the practice of medicine to the discipline of statistics. The Swiss mathematician Daniel Bernoulli even analyzed smallpox mortality to estimate the risks and benefits of inoculation (the fatality rate among those inoculated exceeded the benefit in population survival). And John Snow famously established the route of transmission for

---

* Mortality tables aren't always as accurate as they are precise. A table showing the reasons for 942 recorded deaths by "Diseases and Casualties" in the city of Boston during the year 1811 lists such causes as "drinking cold water" (2 deaths), "sudden death" (25), "white swelling," and "decay." To be sure, the most common was "consumption"—tuberculosis, which was tagged with 221 of the total, or 23 percent of all deaths. Even as late as 1900, tuberculosis was responsible for 194 out of every 1,000 deaths in the United States. See Chapter Six.

London's nineteenth-century cholera epidemics, tracing them to a source of contaminated water.

But plotting the disease pathways, and even recording the traffic along them, did nothing to identify the travelers themselves: the causes of disease. More than a century after the Dutch draper and lens grinder Anton van Leeuwenhoek first described the tiny organisms visible in his rudimentary microscope as "animalcules" and the Danish scientist Otto Friedrich Muller used the binomial categories of Carolus Linnaeus to name them, no one had yet made the connection between the tiny creatures and disease.

The search, however, was about to take a different turn. A little more than a year after Napoleon's death in 1821, a boy was born in the France he had ruled for more than a decade. The boy's family, only four generations removed from serfdom, were the Pasteurs, and the boy was named Louis.

The building on rue du Docteur Roux in Paris's 15th arrondissement is constructed in the architectural style known as Henri IV: a steeply pitched blue slate roof with narrow dormers, walls of pale red brick with stone quoins, square pillars, and a white stone foundation. It was the original site, and is still a working part of one of the world's preeminent research laboratories: the Institut Pasteur, whose eponymous founder opened its doors in 1888. As much as anyone on earth, he could—and did—claim the honor of discovering the germ theory of disease and founding the new science of microbiology.

Louis Pasteur was born to a family of tanners working in the winemaking town of Arbois, surrounded by the sights and smells of two ancient crafts whose processes depended on the chemical interactions between microorganisms and macroorganisms—between microbes, plants, and animals. Tanners and vintners perform their magic with hides and grapes through the processes of putrefaction and fermentation, whose complicity in virtually every aspect of food production, from pickling vegetables to aging cheese, would fascinate Pasteur long before he turned his attention to medicine.

For his first twenty-six years, Pasteur's education and career followed the conventional steps for a lower-class boy from the provinces heading toward middle-class respectability: He graduated from Paris's École Normale Supérieure, then undertook a variety of teaching positions in Strasbourg, Ardèche, Paris, and Dijon. In 1848, however, the young teacher's path took a different turn—as, indeed, did his nation's. The antimonarchial revolutions that convulsed all of Europe during that remarkable year affected nearly everyone, though not in the way that the revolutionaries had hoped.

Alexis de Tocqueville described the 1848 conflict as occurring in a "society [that] was cut in two: Those who had nothing united in common envy, and those who had anything united in common terror." It seems not to have occurred to the French revolutionaries that replaced the Bourbon monarchy with France's Second Republic that electing Napoleon Bonaparte's nephew as the Republic's first president might not work out as intended. Within four years, Louis-Napoleon replaced the Second Republic with the Second Empire . . . and promoted himself from president to emperor.

France's aristocrats had cause for celebration, but so, too, did her scientists. The new emperor, like his uncle, was an avid patron of technology, engineering, and science. Pasteur's demonstration of a process for transforming "racemic acid," an equal mixture of right- and left-hand isomers of tartaric acid, into its constituent parts—a difficult but industrially useless process—both won him the red ribbon of the Légion d'honneur, and earned him the attention of France's new leader. The newly crowned emperor Napoleon III was a generous enough patron that, by 1854, Pasteur was dean of the faculty of sciences at Lille University—significantly, in the city known as the "Manchester of France," located at the heart of France's Industrial Revolution. And the emperor did him another, perhaps more important, service by introducing the schoolteacher-turned-researcher to an astronomer and mathematician named Jean-Baptiste Biot. Biot would be an enormously valuable mentor to Pasteur, never more so than when he advised his protégé to investigate what seemed to be one of the secrets of life: fermentation.

Credit: National Institutes of Health/National Library of Medicine

Louis Pasteur, 1822–1895

At the time Pasteur embarked on his fermentation research, the scientific world was evenly divided over the nature of the process by which, for example, grape juice was transformed into wine. On one side were advocates for a purely chemical mechanism, one that didn't require the presence or involvement of living things. On the other were champions for the biological position, which maintained that fermentation was a completely organic process. The dispute embraced not just fermentation, in which sugars are transformed into simpler compounds like carboxylic acids or alcohol, but the related process of putrefaction, the rotting and swelling of a dead body as a result of the dismantling of proteins.

The processes, although distinct, had always seemed to have something significant in common. Both are, not to put too fine a point on it, aromatic; the smell of rotten milk or cheese is due to the presence of butyric acid (which also gives vomit its distinctive smell), while the smells of rotting flesh come from the chemical process that turns amino acids into the simple organic compounds known as amines, in this case, the aptly named cadaverine and putrescine, which were finally isolated in 1885. But did they share a cause? And if so, what was it? The only candidates were nonlife and life: chemistry or biology.

The first chemical analysis of fermentation—from sugar into alcohol—was performed in 1798 by the French polymath Antoine Lavoisier, who called the process "one of the most extraordinary in chemistry." Lavoisier described how sugar is converted into "carbonic acid gas"—that is, CO_2—and what was then known as "spirit of wine" (though he wrote that the latter should be "more appropriately called by the Arabic word *alcohol* since it is formed from cider or fermented sugar as well as wine"). In fact, the commercial importance of all the products formed from fermentation—wine, beer, and cheese, to name only a few—was so great that, in 1803, the Institut de France offered the prize of a kilogram of gold for describing the characteristics of things that undergo fermentation. By 1810, in another industrial innovation, French food manufacturers figured out how to preserve their products by putting them into closed vessels they then heated to combust any oxygen trapped inside (and, hence, inaugurating the canning industry). Since the oxygen-free environment retarded fermentation, and, therefore, spoilage, it was believed that fermentation was somehow related to the presence of oxygen: simple chemistry.

However, another industrial innovation, this time in the manufacture of optical microscopes, provided another theory. In the 1830s, the Italian astronomer Giovanni Amici discovered how to make lenses that magnified objects more than five hundred times, which allowed observers to view objects no wider than a single micron: a thousandth of a millimeter. The first objects examined were the ones most associated with commercially important fermentation: yeasts.* In 1837, the German scientist Theodor Schwann looked through Amici's lenses and concluded that yeasts were, in fact, living things.

As with many such breakthroughs, Schwann's findings didn't convince everyone. To many, including Germany's preeminent chemist, Justus von Liebig, this smacked of a primitive form of vitalism. It seemed both simpler, and more scientific, to attribute fermentation to a simple interaction of exposing sugar to air. The battle would go on for decades,† until Pasteur summed up a series of experiments with what was, for him, a modest conclusion. "I do not think," he wrote, "there is ever alcoholic fermentation unless there is simultaneous organization, development, and multiplication of" microscopic animals. By 1860, he had demonstrated that fermenting microorganisms were responsible for spoilage—turning milk sour, and grape juice into wine. And by 1866, Pasteur, by then professor of geology, physics, and chemistry in their application to the fine arts at the École des Beaux Arts, published, in his *Studies on Wine*, a method for destroying the microorganisms responsible for spoiling wine (and, therefore, milk) by heating to subboiling temperatures—60°C. or so—using a process still known as pasteurization. He even achieved some success in solving the problem of a disease that was attacking silkworms and, therefore, putting at risk France's silk industry.

The significance of these achievements is not merely that they provide evidence for Pasteur's remarkable productivity. More important, they were, each of them, a reminder of the changing nature of science itself. In an era when national wealth was, more and more, driven by technological prowess rather than the acreage of

---

* Another reason is that yeasts, which are multicellular fungi, are much larger than single-celled bacteria, which had not yet been reliably identified.
† When Schwann published his results, Liebig responded in what might seem, in retrospect, a slightly juvenile manner. In 1839, he wrote—anonymously—a paper in which "the riddle of alcoholic fermentation [is] solved" by a tiny animal shaped like a still, which swallowed sugar at one end and excreted alcohol from its anus and carbonic acid from its penis.

land under cultivation, the number of laborers available, or even the pursuit of trade, industrial chemistry was a strategic asset. France was Europe's largest producer of wine and dairy products, and the weaver of a significant amount of the world's silk, and anything that threatened any of these "industries" had the attention of the national government.

The next twenty years would revolutionize medicine further still, and once again Pasteur would be at the revolution's red-hot center, establishing the critical connection between fermentation, putrefaction, and disease.

As early as 1857, Pasteur had disputed Liebig's position that putrefaction was the cause of fermentation and contagious disease—that both were, in some sense, a result of rot. The leap was to invert Liebig's logic. When Pasteur had examined the diseased silkworms in Lille, and the ailing wine grapes he had studied in Arbois, what he saw in the microscope looked exactly like fermentation. Since he knew that fermentation was caused by microorganisms like yeasts and even smaller living things, fermentation and disease must have a common, microbial cause.

Pasteur wasn't the only one to arrive at this hypothesis.

The Frenchman's first notable breakthrough, in the revolutionary year of 1848, was the discovery that two molecules can be made up of the same ingredients but structurally be mirror images of one another. One of the by-products of wine fermentation, tartaric acid, which is composed of four carbon atoms, six hydrogen, and six oxygen, is a "right-handed" molecule: Light passing through it is bent in a right-handed direction. Racemic acid, on the other hand, has the same formula—$C_4H_6O_6$—but rotates light in both directions: It is, in formal terms, both *dextrorotatory* and *levororotatory*. This discovery was important enough on its own terms, as any high school chemistry student who has struggled with the concept of stereochemistry can testify. Rotating the lens of history on Pasteur reveals his only serious rival for the title of "Father of the Germ Theory of Disease" (also "Father of Microbiology") was Robert Koch: his chiral double.

Robert Koch, 1843–1910

Credit: National Institutes of Health/National Library of Medicine

Koch was born in 1843 in the town of Claustal in the Kingdom of Hanover, one of the principalities that preceded the creation of the modern German state. Like Pasteur, he was a beneficiary of an entire nation's newfound enthusiasm for technical education, even more profound in the Germanophile world than in France. This was especially true in medical education; every major hospital in the patchwork of German-speaking states had been aligned with a university since the late eighteenth century, when Joseph Andreas von Stifft opened the Vienna Allgemeine Krankenhaus as part of the University of Vienna.

In 1844, a year after the birth of Koch, Carl von Rokitansky succeeded Stifft, and the "Vienna School of Medicine" linked examination of living patients with the results of autopsies performed on the same patient. By separating clinical medicine from pathology—before Rokitansky the same doctors had been responsible for both—and documenting more than sixty thousand autopsies, they built a huge diagnostic database that could be validated by postmortem studies.

As a result of the German-speaking world's embrace of scientific education, especially in medicine, Koch attended at least as rigorous a secondary school and university as Pasteur, where, again like Pasteur, he acquired both a medical degree and a mentor who planted the seed for his future researches. For Koch, that mentor was Jacob Henle, professor of anatomy at the University of Göttingen, who had been an advocate for the idea of infection by living organisms since the 1840s.

Both Pasteur and Koch were gifted experimentalists, happiest working in laboratories.

Pasteur came to the experiments that would revolutionize medicine by a relatively roundabout route: first studying the basic chemistry of organic molecules, then the phenomenon of industrial fermentation. Koch did so more directly. As medical officer for the Rhineland town of Wöllstein, he began studying a disease then decimating herds of farm animals in his rural district.

Anthrax was and remains a deadly disease both to all sorts of herbivorous animals who contract it while grazing, and (rarely) to carnivores who get it secondhand from their prey. Every year of the nineteenth century, it killed hundreds of thousands of European cows, goats, and sheep. It was also a feared killer of the humans who acquired it indirectly from infected animals they handled as herders or ranchers; a particularly deadly form is known as "wool-sorter's disease," for obvious reasons. For both animals and human victims, anthrax kills in a gruesome fashion: Lethal toxins* cause severe

---

* One of the three components of anthrax toxin—somewhat strangely, none of the three are toxic in themselves, but only in combination—is actually called "lethal factor endopeptidase."

breathing problems, the painful tissue swelling known as edema, and eventually death. Koch was determined to learn the disease's secrets. What caused it? How was it transmitted from sick to healthy organisms? Most important: Could it be prevented or cured?

Even by the standards of the day, Koch's lab equipment was notably primitive; he inoculated twenty generations of lab mice with fluids taken from the spleens of dead cows and sheep, using slivers of wood. It was an object lesson in the importance of rigorous technique rather than sophisticated tools. Koch's wood slivers established that the blood of infected animals remained contagious even after the host's death. They showed him that it wasn't the blood itself, but something within the blood, that carried the disease. In order to find it, he needed a pure sample of the contagious element. Once again working with home-made equipment, he isolated and purified the element and caused it to multiply in a distinctive environment: the watery substance taken from inside the eye of an uninfected ox, where he cultivated pure cultures of it. When he injected the cultured fluid into healthy animals, they contracted anthrax.

He had his pathogen. The country doctor had forged, for the first time, a link between a distinct microorganism and a single disease. And he had something more, something that explained how grazing animals contracted the disease that he was able to grow only in very specific conditions. When unable to grow, anthrax produces spores that allow it to survive in the absence of food, a host, or even oxygen (in, for example, the well-ploughed soil of Wöllstein). When conditions improve—after they enter either the digestive or respiratory system of some unlucky bovine—they germinate and start multiplying again. Soon enough, the toxins that cause the disease reach deadly levels.

This was big enough news that it attracted the attention of Ferdinand Julius Cohn, a professor of botany at the University of Breslau, and, perhaps most notably, the scientist who had, in 1872, named an entirely new class of living thing: *bacteria* (from *baktron*, the Greek word for "staff"). By then, the complicity of the tiny organisms that Leeuwenhoek had called animalcules in both fermentation and disease was starting to become doctrine. Awareness of the existence of a connection between disease and bacteria didn't, however, tell much about the microorganism's pervasiveness, the mechanism by which they killed, or even how long they had existed.

One reason for the nineteenth century's ignorance about the age of bacteria was a comparable deficiency in knowledge about the age of the earth itself. Until the beginning of the twentieth century, the oldest estimates for the planet's origins were those of William Thomson, Lord Kelvin, who had used the equations of thermodynamics to calculate that the earth was approximately twenty million years old. This was a massive problem for advocates of Darwinian evolution, including Darwin himself, who was "greatly troubled" by it, since a mere twenty million years were insufficient for anything like the known diversity of life on earth.

Current estimates of the age of the planet, roughly 4.6 billion years, solve that problem. For most of that unimaginably long time, bacteria were the dominant form of life on earth. By some estimates, they still are. Recognizable bacterial life—single celled, with a full suite of metabolic tools, but without a nucleus—first appeared about 3 billion years ago, and were the *only* form of life on earth until about 570 million years ago. They remain by far the most numerous. A single gram of topsoil can contain more than forty million bacteria; an ounce of seawater thirty million. Overall, the mass of the earth's 5×10^{30} bacteria may well exceed that of all the plants and animals combined.

Until the middle of the twentieth century, bacteria were a puzzle for taxonomists, who had spent centuries assuming that all living things were either plants or animals. It wasn't until the 1930s that a French marine biologist named Edouard Chatton came up with a different, and more accurate, bifurcation of the living world, dividing it between organisms that possess, and those that lack, a cellular nucleus, the "kernel of life." The Greek word for "kernel," *karys*, gave Chatton his naming convention: Bacteria are prokaryotes (in French, *procariotique*); virtually everything else, from Pasteur's yeasts to a

blue whale, is a eukaryote.* Bacteria live everywhere from Arctic gla-
ciers to superheated vents in the ocean floor to mammalian digestive
systems. They have been around so long—the number of generations
that separate the first bacteria from the ones that probably gave George
Washington his sore throat is about 3×10^{11}, roughly six orders of
magnitude greater than the number of generations separating the
general from the first yeasts—that they have become past masters of
evolutionary innovation. Bacteria, which reproduce as frequently as
three times an hour, can mutate into entirely new versions of themselves in what is, to every other organism on the planet, a figurative
blink of an eye. And, perhaps most relevant for a history of disease,
they can feed themselves on everything from sunlight, to chemicals
so toxic that they are used to clean the undersides of ships, to, well, us.
In the words of one twentieth-century biologist, "It is not surprising
that microbes now find us so attractive. Because the carbon-hydrogen
compounds of all organisms are already in an ordered state, the human body is a desirable food source for these tiny life forms."

Though its age and extent was unknown to Cohn, he did know
that the microorganism that Koch had found was part of this bacterial
universe. He published Koch's work in his journal, *Beiträge zur Bi-
ologie der Pflanzen*—in English, *Contributions on Plant Biology*—
in 1876. The discovery immediately turned Koch into one of Europe's
best-known life scientists. Which brought him to the attention of an
even more famous one: Louis Pasteur.

In 1877, Pasteur took it upon himself to resolve what remained of
the debate about the causes of anthrax. The bacteria isolated by Koch
were still thought to be, in the words of at least one biologist, "neither
the cause nor necessary effect of splenic fever [i.e., anthrax]" since

* As of this writing, the most widely accepted tree of life has two domains: prokarya,
which includes both the kingdoms of the bacteria, and microoganisms that live in
extreme environments that resemble that of the young and oxygen-free earth,
known as archaea; and the eukarya, with four kingdoms: plantae, animalia, fungi
(single-celled organisms, like the yeasts), and protista (many algae, also single-celled
microbes).

exposure to oxygen destroyed them, but material containing the dead organisms still caused anthrax. Pasteur wasn't convinced. He repeated the same process used by Koch: successive dilution—essentially taking a few drops from a flask in which he grew anthrax, diluting it in a new flask, over and over, until there was no doubt that every other potentially contagious element had disappeared—then injecting the pure culture into host animals, who reliably contracted the disease. The endospores discovered by Koch were the reason that seemingly dead bacteria remained carriers of disease: Anthrax cells weren't killed by oxygen, but simply became dormant inside the walls of a spore. The groundbreaking understanding of anthrax—a combination of Koch's spores and Pasteur's dilutions—was the first to link the German physician with the French chemist. It would not be the last.

The next step for Pasteur was the transformation of his experimental results into a practical therapy. A hundred years before, the English physician Edward Jenner had demonstrated that exposing healthy subjects to fluid taken from (relatively benign) cowpox lesions conferred immunity to smallpox. Pasteur himself had discovered that injecting hens with cultures containing chicken cholera that had lost virulence offered the same protection against that disease. Why not anthrax? The key, as always in vaccination, was to find a version of the disease-causing agent that was weakened enough that exposure to it was unlikely to cause the disease itself. In 1881, Pasteur and several of his colleagues, including Charles Chamberland and Emile Roux, used a variety of methods, such as exposure to acids or to different levels of heat, to reduce the disease-causing powers of the bacterium, though not without complications. Pasteur's absolute belief in his irreplaceable gifts had by then taken the form of insufferable arrogance; in 1878, he had attacked the brilliant physiologist Claude Bernard, who questioned the argument that fermentation required living organisms, as a near-blind, publicity-seeking fraud.* And so, naturally, when the veterinarian Jean-Joseph Henri

---

* It wasn't until Pasteur's death in 1895 that Eduard Buchner announced the key discovery that proved both Bernard and Pasteur correct: Enzymes—products

Toussaint successfully attenuated the anthrax bacterium first, Pasteur was furious. So much so that when his team discovered a procedure that weakened the bacteria sufficiently to make a practical vaccine using Bernard's technique—potassium dichromate, an oxidant—he claimed to have done so by using oxygen alone in order to avoid sharing credit.

His tactics worked, both as a disease preventative and as a public relations coup. On May 5, 1881, at the French village of Pouilly-le-Fort, fifty sheep and ten cows were divided into a control group and an experimental one, with the latter receiving the vaccination, and all subsequently exposed to anthrax bacilli. A month later, all the unvaccinated animals in the control group had died; none of the vaccinated ones had even contracted the disease. Anthrax had been defeated by Louis Pasteur. Already France's favorite scientific hero, he was now mentioned in the same breath as Lavoisier and Blaise Pascal.

Germany reacted less enthusiastically. After attending Pasteur's 1882 presentation to the Fourth International Congress for Hygiene, Robert Koch wrote a ten-thousand-word-long screed, of which the following offers a taste:

> Pasteur began with impure material, and it is questionable whether inoculations with such material could cause the disease in question. But Pasteur made the results of his experiment even more dubious by inoculating, instead of an animal known to be susceptible to the disease, the first species that came along—the rabbit. . . . Pasteur follows the tactic of communicating only favorable aspects of his experiments, and of ignoring even decisive unfavorable results. Such behavior may be appropriate for commercial advertising, but in science it must be totally rejected. At the beginning of his Geneva lecture, Pasteur placed the words "Nous avons tous une passion supérieure, la passion de vérité." ["We have no passion

and components of the metabolic activity of microorganisms—were what caused fermentation.

*greater than the passion for truth."] Pasteur's tactics cannot
be reconciled with these words. His behavior is simply inex-
plicable. . . .*

Koch's report was nothing less than a declaration of war, one that
would last until Pasteur's death in 1895 (some would say, ending not
even then). By 1880, Koch had moved to a much-improved lab at the
Imperial Health Bureau in Berlin, from which discoveries continued
to appear on what seemed to be a monthly basis. Having learned, by
hard trial and error, that growing bacteria in nutrient-rich liquids like
beef broth was a losing game—colonies of different sorts mixed together far too easily—he discovered that he could grow pure strains
on potato slices, and later on what is still the standard growth medium for bacterial cultures, the seaweed-based jelly known as agar.
His assistant, Julius Richard Petri, designed and built the eponymous
dishes on which agar compounds would host microbial colonies. In
1882, Koch discovered the bacterium that caused tuberculosis, an
achievement that his colleague Friedrich Löffler called a "world-
shaking event" that transformed Koch "overnight into the most successful researcher of all times."* In 1885 he discovered and identified
the bacterium responsible for cholera.

Nor did he limit himself to experimental work in his Berlin lab;
Koch formulated criteria for managing cholera epidemics, and created, with Löffler, what became known as the "four postulates" of
pathology, a diagnostic tool that would link a single pathogen to
a single disease. The postulates themselves were plausible and useful. The first states that a pathogen must be found in all organisms
that are a disease's victims, but not in any healthy organisms. The
second, that the microorganism must be isolated from a diseased organism and multiply in a culture like agar. The third postulate held
that a cultured microbe must continue to cause disease in a healthy

---

* For more about tuberculosis, see Chapter Six.

host. Finally, the fourth required that the reisolated microbe would be the same as the original one.*

It seemed at the time to scientists throughout Europe that it was Pasteur's unwillingness to use the postulates as a diagnostic tool that caused so much of Koch's hostility. No doubt it was a contributing factor. Another was more basic: Koch was German, Pasteur French. And neither had forgotten the year 1870.

In July 1870, the French parlement, in support of Pasteur's patron, the Emperor Napoleon III, voted to declare war on Prussia. They had been maneuvered into doing so by Prussia's prime minister, Otto von Bismarck, whose North German Confederation swiftly and decisively destroyed the French armies in the east, captured the emperor, besieged Paris, accepted France's surrender, and proclaimed a new German Empire under the Prussian king, all in less than a year. This upset the balance of power that had been obtained in Europe after the defeat of Bonaparte, but not as much as it upset Louis Pasteur, whose reaction was the opposite of moderate: "Every one of my works to my dying day will bear the epigraph: *Hatred to Prussia. Vengeance. Vengeance.*" That Koch was German—Hanoverian, not Prussian, not that it mattered—didn't escape Pasteur's notice. Nor the fact that Koch had served as a military surgeon during the war.

So it went, then, for the remainder of the two men's lives: two brilliant experimentalists revolutionizing the practice of biology and medicine, each with a résumé listing a dozen achievements, any one of which would have bought them scientific immortality, each honored in every way that their nations could honor them. And, as if to underline the mirror-image metaphor, each was retrospectively dogged by accusations of what might kindly be called embellishment.

---

* Plausible, but wrong. Many pathogens can be found in otherwise healthy carriers, which violates postulate number one, and even when they can be grown in a Petri dish (and not all disease-causing agents can be cultured) don't cause disease in every new host, which made Koch's tool an exercise in both diagnostic reductionism and the contingent nature of knowledge, since they were upended even during their creator's own life.

For Robert Koch, the moment of overreach wouldn't come until 1890, when, working at Berlin's Imperial Health Bureau, he announced the discovery of a new therapeutic technique for tuberculosis, an extract of the tubercle bacillus that he named "tuberculin." By then, Koch's reputation was sufficiently great that his word was sufficient grounds for widespread acceptance; a cure for one of the most dangerous diseases known was at hand, and it was used as a treatment over the course of the next eleven years. That it took so long before tuberculin's therapeutic uselessness was discovered is partly because the subjects on whom it was used were already so sick that frequent failure was expected, even forgiven. Koch himself was not so easily forgiven. He had kept his formulation a secret, which was damning enough, but his reason was worse: He had made no secret of his intention of profiting by it, and was therefore unwilling to share potentially valuable trade intelligence with other scientists. Moreover, when he was finally forced, by reports that the substance was actually harmful (as it turned out, the bacteria, even in glycerin, produced the equivalent of an allergic reaction),* it became clear that Koch had only the sketchiest idea of its ingredients, nor could he provide the guinea pigs that he had supposedly cured with it. It took until the end of Koch's life for his reputation to recover from the scandal, a testimony both to the limitations of the vaccine approach itself—as we shall see, *attacking and defeating* infectious disease is a very different process than *defending against* it—and to the pressures of scientific discovery.

Pasteur's reputation managed to remain largely unsullied by any similar scandal, at least during his lifetime. It was only a century later that one of his most widely publicized achievements, his rabies vaccine of 1885, was revealed, in a biography entitled *The Private Science of Louis Pasteur*, to be considerably less significant than it had appeared.

By 1885, the search for a bacterium that caused rabies had failed—inevitably, since the disease is caused by a virus, one of those freefloating bits of genetic material wrapped in a protein coat that are so much smaller than bacteria that they reproduce inside them, and that

* This is the basis of the modern TB skin test.

wouldn't even be identified until 1892. However, Pasteur's thinking went, since rabies had a very slow gestation period—anywhere from a month to a year—perhaps a vaccine could actually "cure" the disease, by immunizing the victim from it *after* infection, but *before* symptoms appeared. That is, a vaccination given after exposure to the bite of a rabid animal could serve to inoculate the victim against the disease, which was inevitably fatal once symptoms appeared.

So, when nine-year-old Joseph Meister was bitten by a rabid dog in July 1885, and survived after Pasteur inoculated him with a weakened rabies virus that had been cultivated in live rabbits, the public acclaim was enormous. The French public practically deified Pasteur, and raised more than 2.5 million francs—at least 12 million dollars today—that enabled the Institut Pasteur to open its doors three years later.

Meister made Pasteur a national hero, though at least partly through a misunderstanding of just what "curing" a disease means. A bite from a rabid dog will cause rabies in a human only about one time in seven (though, in 1885, that one would invariably die). Had Pasteur not inoculated Meister, he had a good chance of surviving anyway. Another, rather larger problem with the story of Pasteur's heroic achievement is that Meister wasn't the first victim to receive Pasteur's vaccine; two weeks before, in June 1885, a girl named Julie-Antoinette Poughon had been given the vaccine but died shortly thereafter. Nor had Pasteur tested the method on dogs prior to giving it to Meister, though he claimed to have done so. Pasteur, perhaps understandably, neglected to mention either fact to journalists or other scientists.*

For Koch and Pasteur, however, the real achievements remain so outsized that embarrassments that would destroy the reputation of garden-variety scientists seem barely to rise to the level of peccadillo. Pasteur, especially, was a hero from the time of his original fermentation discoveries, and not just in France. In 1867, the Englishman Joseph Lister was so taken with Pasteur's researches that he wrote:

---

* In any case, rabies was scarcely a large public health hazard. In 1885 France, fewer than thirty people died of it annually.

Turning now to the question how the atmosphere produces decomposition of organic substances, we find that a flood of light has been thrown upon this most important subject by the philosophic researches of M. Pasteur, who has demonstrated by thoroughly convincing evidence that it is not to its oxygen or to any of its gaseous constituents that the air owes this property, but to the minute particles suspended in it, which are the germs of various low forms of life, long since revealed by the microscope, and regarded as merely accidental concomitants to putrescence, but now shown by Pasteur to be its essential cause, resolving the complex organic compounds into substances of simpler chemical constitution, just as the yeast plant converts sugar into alcohol and carbonic acid.

When he wrote this, Lister was a working physician and the Regius Professor of Surgery at the University of Glasgow. He had been born in Essex, forty years before, to a prosperous and accomplished Quaker family. His father, Joseph Jackson Lister, was a respected physicist and pioneer of microscopy; a classic scientific amateur, he was elected a Fellow of the Royal Society, the world's oldest and most respected scientific organization, for inventing the achromatic microscope.

In 1847, Joseph graduated from University College London—even in the middle of the nineteenth century, both Oxford and Cambridge were still barred to Quakers—and entered the Royal College of Surgeons. In 1853, he became a house surgeon at University College Hospital, and three years later was appointed surgeon to the Edinburgh Royal Infirmary.

In 1859, the newly married Lister moved to the University of Glasgow, and his story really began.

Credit: National Institutes of Health/National Library of Medicine

Joseph Lister, 1827–1912

The Glasgow Royal Infirmary had been built in the hope that it would prevent "hospital disease" (the name coined by the Edinburgh obstetrician James Young Simpson in 1869 for the phenomenon known today as "surgical sepsis" or "postoperative sepsis"). In this, it was a notable failure. In Lister's own records of amputations performed at the Glasgow Royal Infirmary, "hospitalism" killed between 45 and 50 percent of his patients. Lister wrote, "Applying [Pasteur's] principles to the treatment of compound fracture, bearing in mind that it is from the vitality of the atmospheric particles that all mischief arises, it appears that all that is requisite is to dress the wound with some material capable of killing those septic germs, provided that any substance can be found reliable for this purpose, yet not too potent as a caustic."

Lister's earlier work—on the coagulation of the blood, and espe-cially the different microscopically visible stages of inflammation in the sick—convinced him that Pasteur had it right. The "pollen" idea, how-ever, convinced him also that the microorganisms traveled exclusively through the air. This was wrong, but usefully so, since it argued for the most impassable barrier between the "infected" air and the patient.

Not a physical barrier, a chemical one. In 1834, the German chem-ist Friedlieb Runge had discovered that what he called *Karbolsäure* could be distilled from the tarry substance left behind when wood or coal are burned in furnaces or chimneys: creosote, the stuff that gives smoked meat its flavor. Sometime in the 1860s, Lister read an article about how a German town used creosote to eliminate the smell of sewage. Since he knew, pace Pasteur, that the smell of sewage was caused by the same chemical process that caused wounds to mortify, he reasoned that a compound that prevented one might, mutatis mu-tandis, halt the other. In the spring of 1865, he started testing other coal tar extracts on patients, and, on August 12, he hit the jackpot: The substance known as phenol, or carbolic acid, stopped infections cold.* Two years later, he published his results: Surgical mortality at the Glasgow Infirmary had fallen from 45 percent to 15 percent. "As there appears to be no doubt regarding the cause of this change, the importance of the fact can hardly be exaggerated."†

It took years before Lister was able to persuade the medical estab-lishment of the importance of what has come to be known as antisep-sis, helped along more by practical and highly publicized results—in 1871, he safely drained an abscess under the arm of Queen Victoria—

---

* He did not, of course, know how. Phenol causes cell membranes to degrade and eventually rupture, with a resulting—and devastating—leakage of the cell's internal constituents.

† It should be noted that many other factors were simultaneously contributing to a drop in surgical complications, including simple hygiene. Moreover, Lister's view of the causes of infection evolved considerably from 1867, when he still held to the belief that at least some diseases were the result of *miasmata*, or bad air, and seemed to favor the Liebig view that putrefaction was a cause of microbial action, rather than a result of it.

than by experimental validation. Dependence on antisepsis, however, had its own risks. Patients were frequently required to inhale the fumes of burning creosote, which was dangerous enough. Even worse, some were given injections of carbolic acid, which doesn't kill just dangerous pathogens, but often enough, the patients themselves. As the German physiologist (and winner of the very first Nobel Prize in Medicine) Emil Behring pointed out in the 1880s, "It can be regarded almost as a law that the tissue cells of man and animal are many times more susceptible to the poisonous effects of disinfectants than any bacteria known at present. Before the antiseptic has a chance either to kill or inhibit the growth of the bacteria in the blood or in the organs of the body, the infected animal itself will be killed."

Lister's reputation, and the importance of both antiseptic *and* aseptic surgical practice—not merely disinfecting wounds, which Lister pioneered, but maintaining a fully sanitary environment around patients, a technique he adopted far later—continued to grow over the rest of his life. He would become one of the heroes of nineteenth-century Britain, president of the Royal Society, founder of the British Institute of Preventive Medicine (renamed the Lister Institute of Preventive Medicine in 1903), and be made Baron Lister of Lyme Regis. In 1899, the Chinese minister to the Court of St. James's, commanded by the emperor to produce biographies of the hundred greatest men in the world, announced that the three Englishmen to make the cut were William Shakespeare, William Harvey, and Lister himself. In retrospect, this seems modest enough. The germ theory of disease that had been developed and tested by Pasteur, Koch, and Lister himself produced an astonishing number of discoveries about the causes of disease; not merely anthrax, tuberculosis, and cholera—respectively the bacteria known as *Bacillus anthracis, Mycobacterium tuberculosis,* and *Vibrio cholerae*— but gonorrhea (*Neisseria gonorrhoeae,* discovered 1879), diphtheria (*Corynebacterium diphtheriae,* discovered 1883), bacterial pneumonia (*Streptococcus pneumoniae,* discovered 1886), gas gangrene (*Clostridium perfringens,* discovered 1892), bubonic plague (*Yersinia pestis,* discovered 1894), dysentery (*Shigella dysenteriae,* discovered 1898), syphilis (*Treponema pallidum,* discovered 1903), and whooping cough

(*Bordatella pertussis,* discovered 1906). Moreover, the discovery of the nature of these infectious agents led directly to a powerful suite of defensive weapons; not merely antisepsis and vaccination, but even more usefully, improved sanitation and hygiene. Since by their very nature, preventive measures succeed when disease doesn't even appear, it is impossible to know with certainty how many lives were saved by these expedients, but they are the most important reason that European life expectancy at birth increased from less than forty years in 1850 to more than fifty by 1900.

Nonetheless, as valuable as these practices were in defending human life from pathogens, millions continued to fall ill from infectious disease every day. And when they did, medicine could do virtually nothing about it. Perversely, the greatest triumph in medical history—the germ theory of disease—destroyed the ideal of heroic medicine, replacing it with a kind of therapeutic fatalism.* As physicians were taught the bacterial causes of diseases, they also learned that there was little if nothing to do once a patient acquired one.

In one of Aesop's best-known fables, a group of frogs living in a pond prayed to the gods to send them a king; an amused Zeus dropped a log in the pond, and announced that this was, henceforth, the frogs' king. The frogs, disappointed with their new king's inactivity, prayed again for a king . . . this time one that would *do* something, upon which Zeus sent them a stork, who promptly ate the frogs. The Aesopian moral—always choose King Log over King Stork—is one that the Western world's physicians took to heart, and from the 1860s to at least the 1920s, humility reigned. Only a few drugs had any utility at all (mostly for relieving pain), which made for skepticism about virtually all treatment. On May 30, 1860, Dr. Oliver Wendell Holmes, Sr., famously announced in an address before the Massachusetts Medical Society:

---

* The term used by most historians of the phenomenon is actually "therapeutic nihilism." In fact, the modern version of the Hippocratic Oath includes the phrase, "I will apply for the benefit of the sick, all measures [that] are required, avoiding those twin traps of overtreatment and therapeutic nihilism."

Throw out opium, which the Creator himself seems to pre-scribe, for we often see the scarlet poppy growing in the corn-fields, as if it were foreseen that wherever there is hunger to be fed there must also be a pain to be soothed; throw out a few specifics which our art did not discover, and it is hardly needed to apply; throw out wine, which is a food, and the vapors which produce the miracle of anaesthesia, and I firmly believe that if the whole materia medica [medical drugs], as now used, could be sunk to the bottom of the sea, it would be all the better for mankind,—and all the worse for the fishes.

Holmes overstates, but not by much. The achievements of the nineteenth century in revolutionizing medical therapeutics are nothing to sneeze at, including the recognition that sneezing itself was a powerful source of dozens of infectious diseases. The great biologists of the era established a robust theory about disease, along with powerful tools for defending against it, and left behind a model for research, experimentation, and validation.

It takes nothing away from the extraordinary discoveries of Pasteur, Koch, Lister, and others to wonder whether their most enduring contribution to the revolution in medicine wasn't informational but institutional: the modern biological research laboratories. The Institut Pasteur was founded in 1888; the Lister Institute of Preventive Medicine was established in 1891, the same year that the Robert Koch Institute was founded, originally as the Royal Prussian Institute for Infectious Diseases. In 1890, the Royal Colleges of Surgeons and Physicians opened its first research laboratory in London. These establishments weren't only fertile schools for training for the next generations of researchers, or structures in which the best biologists and physiologists in the world could cooperate—and, truth be told, compete—one with the other. They were also magnets for the resources that research demanded—magnets for the philanthropy of wealthy families, and subsidies from national governments. As the nineteenth century turned into the twentieth, the life sciences had not yet become the enormously expensive proposition they would become decades hence. Nonetheless, even frugal research still cost money, and institutional laboratories were, for a time, the most productive place to spend it.

But for the next chapter in the story that leads from George Washington's sickbed to the maternity wing of New Haven Hospital in 1942, there was an even more important development: the marriage between industrial chemistry and medicine.

TWO

"Patience, Skill, Luck, and Money"

A nineteenth-century German opera, *Der Freischütz*—in English, *The Marksman*—tells the story of a young forester who must pass a test of shooting skill to win the heart of his true love. After losing a match to a young peasant, he is persuaded to improve the odds by using *Freikugel* or *Zauberkugel,* an enchanted bullet that could not fail to hit its target.* The magic bullets—the first six under the control of the marksman, the seventh owned by the devil—are a constant presence in collections of European folktales and a familiar trope in nineteenth-century drama.

Even more durably, they inspired the best-known modern usage of the term, which has little to do with the devil or folk mythology. In 1907, "magic bullets" were the theme of the Harben Lectures given at what was then known as Britain's Royal Institute of Public Health. The lecturer, who used the term to describe a targeted drug, one that would attack a disease-causing microbe without harming the host suffering from the disease, was a German physician named Paul Ehrlich.

Ehrlich was then fifty-two years old, and one of the most respected physicians and scientists in the world. Like Robert Koch, he was a product of a terrifyingly rigorous, but undeniably effective, German education. Beginning with the reforms of Wilhelm von Humboldt in Prussia in the early nineteenth century, Germany had taken a far more pragmatic view of modern subjects such as mathematics and

---

* The word also appears in the great German dictionary assembled by the folklorists Wilhelm and Jakob Grimm, referring to a technique used by the Lapps to poison their targets at a distance.

science than had been the case in the United Kingdom or France, and certainly the United States. By 1872, state-controlled education had added, to the nine years of Latin and Greek provided by the *gymnasia*, the *Realgymnasia*, and *Oberrealschules*, and especially in the *technische Hochschulen* (technical colleges), algebra, chemistry, and physics. After decades of providing its citizens with history's most rigorous educational program, Germany had become the world's leader in virtually every field of scientific research. Moreover, since the original reforms had been explicitly made to support the commercial interests of the German state, there were no qualms about partnerships between schools—secondary and postsecondary—and industry. Paul Ehrlich was a notable beneficiary, as his education took him from the Maria-Magdalenen-Gymnasium in Breslau through universities in

Credit: National Institutes of Health/National Library of Medicine

Paul Ehrlich, 1854–1915

Strasbourg, Freiburg, and Leipzig. In 1878, at the age of twenty-four, he received his doctorate for a dissertation on a subject that would occupy him for years, and turn out to be the first link in a chain that led to the first true antibacterial chemical therapy.

Ehrlich's dissertation was titled "Contributions to the Theory and Practice of Histological Dyes." Histology—the word is taken from a Greek root meaning "something that stands upright" and was adopted by nineteenth-century physiology to mean "tissue"—had just come into its own as a credible discipline. Ever more powerful microscopes had made distinguishing one sort of tissue from another possible, but not so easy. Even with lenses that magnified cells hundreds of times, it was difficult if not impossible to identify distinct cell types, without something to improve the contrast between, for example, different sorts of blood cells. The something was staining: Some chemicals have a special affinity for different cell types, and can turn them a particular color while leaving other similarly shaped cells unchanged. Ehrlich's specific contribution was to use a dye with what he called "an absolutely characteristic behavior toward the protoplasmic deposits of certain cells" in blood plasma. He named the "certain cells" *mastzelle*, from the German word for "fattening," because he believed them involved in the process by which cells fed themselves. In this he was wrong—mast cells are part of the immune system (about which more below)—but the real triumph was the discovery that stains could differentiate components of blood: leukocytes, lymphocytes, red blood cells, and so on. The newly minted doctor, who had become known to his classmates as "the man with blue, yellow, red, and green fingers," had made a huge discovery: Different stains were absorbed by different cell types.

The significance to medicine wasn't merely that a new tool for studying human and animal cells had been discovered. The tool could identify *all* kinds of cells, including bacteria. And, so the logic went, if a dye could recognize a specific class of bacterium, could it not also deliver a specific attack on the pathogenic ones? It was not a giant leap to imagine transforming a blob of paint that unerringly marked a dangerous trap into a compound—a magic bullet, perhaps—that would destroy it. Nor was it especially difficult to find the best place to start the search: wherever Robert Koch was working.* In 1891, Ehrlich joined Robert Koch at his Berlin Institute for Infectious Diseases.

The first step was understanding the magic bullets already present in humans and animals: the immune system. Even before Koch and Pasteur had demonstrated diseases were caused by microorganisms, the concept of immunity had been recognized and even classified. Victims of diseases like smallpox who survived were known to be immune to the disease thereafter. Thucydides, writing about the Plague of Athens, recognized that Athenians who had recovered from one encounter with the disease were protected against a second. The discovery of the germ theory suggested an explanation: Whatever provided immunity was attacking and destroying specific pathogens without harming the host. Could this germ-killing machinery be harnessed as a targeted therapy?

Though Ehrlich and his collaborators didn't yet understand it in detail, the vertebrate immune system is composed of a dizzying array of components, some of them just long chains of amino acids, others complex cells, complete with specialized and deadly organelles. Some parts of the immune system are intelligence analysts, able to identify the nature of attacking pathogens; others are messengers, sending out chemical alarm bells to summon destroyer cells to the site of an invasion. There are even components, like the dendritic cells, that identify antigens, and train other cells, the T-lymphocyte white blood cells, to recognize them the next time they pay the host a visit.

The key categorical distinction used for the immune system is between those components that are *innate*—nonspecific defenses that respond to invaders that the organism has never encountered before—

---

* Six weeks after Koch announced the discovery of the tubercle bacillus, Ehrlich had written a paper on staining it, using a dye known as fuchsine, and then leaching the color out with an acid. Shortly thereafter, he used the "acid-fast" staining method to find the bacillus in his own sputum, and spent two years in southern Europe and Egypt recovering from the disease.

and those that are *specific* (or *adaptive*): the specialized defenses created by organisms only after they have encountered a specific pathogen.
Innate first: When an organism—you, perhaps—encounters a microbe
with evil intent, a dozen different proteins that are always circulating
in the bloodstream send out a chemical alarm that activates another
group of proteins known collectively as the cytokines. These molecules
are folded in distinctive ways, permitting them to attach themselves to
a pathogen, and hang on while sending off *another* alarm, summon-
ing cell-based immune defenders like white blood cells. Meanwhile,
other cell-based defenders, Ehrlich's *mastzelle*, release chemicals like
the histamines, which turn up the body's thermostat, causing what is
known as the inflammatory response: fever.

And that's just the off-the-shelf, generic version of the innate system. Far more powerful are the bespoke defenses of an organism that
has previously been exposed to a particular pathogen and has responded
by tailoring a weapon specifically to destroy it. These specialized forces
include cell-based troops like the B-lymphocytes manufactured in the
bone marrow that create highly specific antibodies that hold on to the
surface of invading cells; the T-lymphocytes that can destroy invaders
by punching a hole in the invader's membrane; and macrophages that
can swallow and destroy them. The specialized immune system is a
double-edged weapon; it depends on a very precise match between the
supply of defenses and the demand for them. The inflammatory response can kill hosts as well as pathogens. Chronic lymphoid leukemia,
one of the most dangerous—and common—cancers does its damage
by overproducing B-cells that accumulate in the bone marrow, and so
crowd out the ones needed to fight infections. Victims frequently die,
not from cancer per se, but from the recurrent infections that their immune systems are no longer able to fight.

When Ehrlich and colleagues, most notably the physician Emil
von Behring and the bacteriologist Hans Buchner, started studying
the immune system in the 1890s, however, this taxonomy of immunity was still decades in the future. Their starting points were far
simpler: Koch and Pasteur's discovery that a specific microbe causes a

specific disease, and that exposure to that microbe conferred immunity thereafter.

The first components of the immune system they discovered were the complex of antibacterial macromolecules found in blood serum, originally named "alexins" by Buchner, but renamed the "complement" by Ehrlich, to reflect his discovery that the thirty or so proteins that comprised it complemented the work of other proteins known as antibodies. If those molecules could be made to appear by exposure to the pathogens, then perhaps they could not only prevent future occurrences of a disease but, like Pasteur's rabies vaccine, treat a new one—acting not as a vaccine, but as an antiserum. Behring started looking.

Born in 1854, Emil Behring (not yet "von") was, like Ehrlich, a man who had come to adulthood after Bismarck's 1871 consolidation of most of the Germanophone world as the Second Reich: Imperial Germany. Unlike the Jewish Ehrlich, he was training for a career in the Church before he transferred to a program in medicine, specifically military medicine. After garrison duty at a number of German army bases, he found his way—again like Ehrlich—to an orbit around one of the two great founders of the science of microbiology, at Koch's Hygiene Institute in 1888, and, after 1890, the Institute for Infectious Diseases.

Also in 1890, Behring and his colleague, a visitor from Japan named Kitasato Shibasaburo, engineered the first real breakthrough in serum therapy. Ever since Jenner, the immune system had been activated by exposing its host to a pathogenic organism. Behring and Shibasaburo discovered that the immune system could be used to cure disease by exposing the host not to a pathogenic organism itself, but to the specific toxin released by most pathogens: the toxin that caused the symptoms of disease.

Their first practical demonstration of this discovery was used to treat a very dangerous disease, indeed: diphtheria. *Corynebacterium diphtheriae* was first identified by the Swiss pathologist Edwin Klebs in 1883 (the toxin produced by the bacteria was first identified in 1888 by Pasteur's colleague Emile Roux and independently by the Swiss physician Alexandre Yersin, who would later give his name to the bacterium that causes bubonic plague: *Yersinia pestis*). Diphtheria was then afflicting more than fifty thousand Germans annually with sore throats, fever, and the disease's characteristic membrane covering the tonsils and pharynx.* Up to five thousand would die. So when Behring and Shibasaburo announced, in 1891, that they were able to cure infected rats, guinea pigs, and rabbits by injecting them with a heat-weakened version of the toxin produced by *C. diphtheria*, it was very big news indeed.

The announcement was made well in advance of a proven therapy. Until 1897, when Ehrlich established a standardized unit for measuring diphtheria toxin, the antitoxin was at best unreliable, at worst dangerous. The immediate significance of the enthusiasm that greeted antiserum therapy wasn't on diphtheria treatment, but on an even more consequential aspect of medical history: It attracted the interest of Germany's industrial chemists. In 1892, Behring signed on to collaborate on the production of a diphtheria antitoxin with the Hoechst chemical company, based in Frankfurt-am-Main.

The company was then barely thirty years old, but had nonetheless seen its name change three times: from Teerfarben Meister, Lucius & Co. to Meister Lucius & Brüning to Farbwerke vorm. Meister Lucius & Brüning AG. By the time they went into business with Behring, the founders had evidently decided that simplicity was the better part of vanity and renamed their company Farbwerke Hoechst AG, since Hoechst (or Höchst) was where their first factory was located. The company's primary business is revealed in its original name: The German *Teer* means "tar." *Farben* translates as "color," or more appropriately, "dye."

That Behring would collaborate with a dye company wasn't exactly unusual. To a very great degree, the chemical industry in the late nineteenth century *was* the dye business: Dyes were by far the largest and most lucrative chemical process yet known, and enormously more profitable than, say, medicine. Vegetable-based dyes

---

* The leatherlike pseudomembrane gave the disease and the bacterium their name; the Greek word for "leather" is *diphtheria*.

extracted from madder root, indigo plants, insects, and even shellfish, which was the source of the ancient dye known as "Tyrian purple," have been used for at least three thousand years and probably longer to color textiles, but it took the scientific advances of the Industrial Revolution to create the first synthetic dyes. Most important of all were the aniline dyes, the organic compounds combined in a chemist's lab.* Beginning in the 1830s, anilines had been extracted from coal tar; the same chemist who had distilled Joseph Lister's carbolic acid from creosote also discovered that calcium hypochlorate could turn coal tar into a spectacular indigo dye. By the 1850s, the English chemist William Henry Perkin had accidentally discovered the first synthetic aniline dye, which he named mauveine. Even better, he had discovered a chemical breakthrough that revealed how to produce it by the carload, and the chemical industry was off to the races. Two other German chemists, Carl Liebermann and Carl Graebe, managed to isolate and synthesize alizarin, the active component in the madder root that had been used as a red dye since ancient Egypt. In 1870, Adolf von Baeyer did the same for indigo.

Since dye companies were always looking for innovative methods of producing color, a mastery of tissue staining was a prized talent. It's therefore no coincidence that Hoechst brought Ehrlich—who had stained his first mast cells with aniline dyes—and Behring together. One of the many talents Ehrlich had cultivated was the ability to visualize the structure of dye molecules in three dimensions. By the end of the 1880s, Ehrlich, despite having no formal training as a chemist, had authored more than forty research papers on the chemistry of dyes and developed a dozen different staining methods using them. Working with Hoechst, and Behring, ought to have been a highly productive business; and it would have been, but for the very different objectives of a chemist working for an industrial lab rather than a hospital or university. Though Ehrlich had taken the precau-

---

* For readers who remember high school chemistry fondly: Anilines are compounds in which a phenyl group, a ring made up of six carbon and five hydrogen atoms, is joined to an amine group, or a nitrogen atom is connected to three other atoms of either carbon or hydrogen.

tion of taking out a basic patent on his method of standardizing diphtheria serum antitoxin, a fair reading of the subsequent events is that Behring managed to void those agreements in order to secure greater profits for himself from the collaboration with Hoechst.

The result was the poisonous transformation of the once close collegiality between two of Robert Koch's greatest protégés into the same sort of lifelong hostility that characterized the relationship between Koch and Pasteur. It wasn't merely the patent dispute. Ehrlich objected even more strongly to Behring's attempt to profit from Koch's tuberculin, producing a version marketed by Hoechst, even though the compound had already been shown to be useless. This was a point that Ehrlich made over and over again. "I must . . . be no longer exposed to Behring's crass egotism and money-grubbing. I am not in the least inclined to . . . be subservient to his business shenanigans. I have no mind whatsoever to convert my Institute into a branch establishment or business venture of Behring's. . . . [He] will now bad-mouth me everywhere, but my conscience is untroubled, and whatever he may be up to doesn't faze me in the least."

In any case, Ehrlich was searching for something bigger than antiserums. Though he joined Berlin's Institute for Serum Research and Serum Testing (a grand name for a lab built in an abandoned bakery) in 1896 and continued his work on toxins when he moved to Frankfurt-am-Main's Institut für Experimentalle Therapie in 1899, he already knew the limitations of antiserum therapy. Behring had shown that diseases could be caused not only by microbes, but also by toxins produced by the microbes, and that the body could produce a therapeutic response once exposed to the toxin. Turning this insight into a therapy, however, proved harder than expected. Antiserum therapy depends on the immune system's ability to recognize a particular toxin: to identify it from its surface appearance. However, while some of the most dangerous toxins—exotoxins—are found on the surface of bacterial invaders, most bacterial diseases are caused by *endo*toxins, which do their damage only when the bacterial cell is ruptured. Since antiserum therapy only worked with exotoxins, it was effective against a limited number of pathogens. If the antiserum gun was firing magic bullets, very few of them were going to find their targets.

This didn't mean that Ehrlich's time in Berlin or Frankfurt was wasted; far from it. In 1897, he developed a truly revolutionary theory about the geometric relationship between an antigen and an antibody: the so-called side-chain theory.

Side-chain theory proposed that the membranes that enclosed the cells of multicellular animals were extremely complex chemical machines, each of whose components—Ehrlich's side chains—has an affinity for a particular nutrient needed by the cell. Normally, each side chain is a kind of combination lock that opens when it encounters a needed protein. If any foreign substances—bacteria, viruses, or toxins—fit the same lock, metabolic activity is blocked. The cell responds by producing similar side chains as replacements, but overdoes it—in Ehrlich's words, "nature is prodigal." The extra side chains are then sloughed off into the cell's surrounding fluid; and each of them, by definition, possesses precisely the shape to combine with the invading antigen. It is as if the cell had produced a finely machined gear that fits perfectly with its pathogenic match: antigens that latch onto the invaders and produce all the actions of the immune response—chemical signaling, inflammation, and everything else. The side-chain theory, and its explication of the immune response, was such an elegant and comprehensive discovery that it would win Ehrlich the 1908 Nobel Prize in Medicine.*

A commonplace observation about the Nobel Prizes says that no important work is ever done by winners after collecting the award. In this as elsewhere, Ehrlich was exceptional. Not only could the physician and scientist have been recognized by the Nobel committee for revolutionary medical discoveries made even before the creation of side-chain theory—histological staining; the investigations of the toxins ricin and abrin, which revealed the timing of the immune

---

* However, it has subsequently been expanded and replaced by a number of new and more finely detailed theories of immunity and immunoregulation.

response—but, as time would reveal, his greatest achievement was still to come. That achievement, the birth of chemotherapy, was, nonetheless, very much of a piece with his very first publication nearly three decades before, on the staining of microscopic structures. Like his *mastzelle*, Ehrlich's magic bullet appeared only in proximity to coal tar–based dyes.

The argument at the core of Ehrlich's Nobel Lecture, entitled "Partial Cell Functions," was that the future of microbiology depended less on observational biology, and more on fundamental chemistry.

Chemistry is one of the younger sciences. If you could pluck Isaac Newton out of the seventeenth century and drop him into a twenty-first-century high school, he could teach at least the first few chapters of a contemporary physics course; the laws of mechanics, for example, are still the ones Newton postulated. For that matter, a second-century mathematician could do the same for a full year of geometry or trigonometry. Even biologists still name species using the Linnaean classifications first published in 1735. In chemistry, though, hardly any idea that appeared before the French Revolution has survived except as an historical oddity, like the theory that chemical activity depended on a combustible substance called "phlogiston."

The discipline began to come together with Antoine Lavoisier's 1789 codification of the twenty-three known elements as substances that could not be broken into smaller components. In 1808, John Dalton in England and Joseph Gay-Lussac in France independently derived similar laws about the constituents of gases that led directly to the theory that all gases—all *everything*—were composed of impossibly small distinct elements known as atoms. The science had, for the first time, a usable conceptual framework.

Elements would be discovered over the course of the nineteenth century, including ones essential to life: sodium and potassium in 1807, calcium in 1808, iodine in 1811. Chemical analysis grew sharper and more precise with the introduction of dozens of apparatus. Since the time of Lavoisier, scientists have used combustion to examine the constituents of organic matter: burning the sample using a hollow glass blowpipe, trapping the carbon dioxide and water vapor produced, and measuring their volume in order to calculate how much hydrogen and carbon the sample originally contained. Gay-Lussac improved on this method by exposing samples to anhydrous calcium carbonate, which likewise trapped water vapor. In 1831, Justus von Liebig (the "Father of the Fertilizer Industry" for his discovery that nitrogen, in the form of ammonia, was the essential element in plant metabolism) invented the five-balled glass vessel he called the "Kaliapparat," which trapped carbon dioxide as it passed through a filter of caustic potash, or potassium hydroxide.

Inevitably, as the tools of analysis improved, the temptation to create chemical compounds, which had been the dream of alchemists since antiquity, grew. The shift from analysis to synthesis—and, not so incidentally, from inorganic to organic chemistry—found its starting point with the German chemist Friedrich Wöhler's 1828 synthesis of urea from simpler components such as cyanic acid and ammonia. In 1845, Hermann Kolbe synthesized acetic acid.

There was, however, one big, unavoidable problem with chemical synthesis: Optical microscopes could make cells an object of study for the biologists like Pasteur and Koch, who used them to establish the precepts of the germ theory. But the molecules that are central to all chemical activity couldn't be seen by even the most powerful lenses.* Because molecules aren't perceptible to the eye, building one in a

---

* Not for quite a while, anyway. The 2014 Nobel Prize in Chemistry was awarded for the "development of super-resolved fluorescence microscopy," a technique for getting around what seemed a fundamental limitation of optical microscopes, which was that they could not, even in theory, produce pictures with a resolution greater than half the wavelength of visible light, about 200 nanometers, or 1/20,000 of a millimeter. This is fine for many biological investigations—most bacteria are about 1,000 nanometers in diameter—but not all. Viruses, for example, can be as small as 20 nanometers in cross-section. Even fluorescence microscopy is too blunt an instrument for molecular investigation. A water molecule is only 1/10 of a nanometer in diameter. For *that* kind of picture, an electron microscope, which diffracts beams of electrons instead of waves of light, and can resolve at the (wait for it . . .) *picometer* scale: 1/1,000 of a nanometer.

laboratory was a little like trying to build a scale model of the Eiffel Tower while wearing a blindfold.

Luckily, it's not quite as daunting as that. There are a number of rules that determine how elements bond to, and react with, one another—the equivalent of understanding which bolt attaches to a particular nut in the overall construction of the Eiffel Tower. The ways in which acidic substances like hydrochloric acid react with alkaline ones, such as lye, were intuited as far back as Lavoisier (although he got almost all the elements involved wrong). In 1857, August Kekulé, a German organic chemist, discovered a concept about chemical structure that first-year chemistry students learn as an atom's valence (Kekulé and his contemporaries called them "affinity units" or "combining power"). Different elements, he proposed, have distinct powers of combination, and can bond (or share an electron) with other elements based on *their* combining powers. If the power was one, as with hydrogen, the atom could form only a single bond; for a combining power of two, such as oxygen, either two single bonds or one double bond; and for a power of three, nitrogen, for example, either three single bonds, a double bond and a single bond, or one triple bond. Twelve years later, a Russian chemist named Dimitri Mendeleev published the first of his periodic tables, then containing sixty-five named elements, organized by atomic weight and valence. Though it wasn't until 1897 that J. J. Thomson discovered the electron (and another nineteen years would pass before their essential role in bonding was recognized), from a practical standpoint, chemists knew the rules of bonding one atom to another.

With the discovery of electrons, though, a huge suite of different chemical reactions could now be explained: reduction reactions, in which electrons are gained, and oxidation reactions, where they are lost; acid-base reactions, mediated by charged particles—*ions*—where positive protons seek out negative electrons in base molecules. As a result, the chemistry that Paul Ehrlich was able to teach himself was sophisticated enough that he could perform experiments on extraordinarily complex chemical structures.

As far back as the mid-1880s, Ehrlich had experimented with the use of azo dyes*—these are aniline derivatives with names like methylene yellow, congo red, and alizarin yellow—as potential therapeutic agents. About 1891, he identified a variant of the dye methylene blue[†] as a treatment for malaria—a distinctly mediocre treatment. First, the compound was only mildly effective; also, due to its origins as a coloring agent, it turned urine green and the whites of the eyes blue, which understandably limited its appeal to both patients and physicians. Even with its problems, though, the dye-based treatment was effective enough that methylene blue pills remained a first-line antimalarial drug well into the 1940s, and promising enough to encourage Ehrlich in the search for his *Zauberkugel*.

He had, by then, learned that testing a magic bullet required a fairly easy-to-hit target, and believed he had found one in trypanosomiasis, the "sleeping sickness" caused by the introduction of trypanosomes into the bloodstream of infected animals, generally by the bite of the tsetse fly. The disease seemed perfect: first, because the trypanosomes that cause it are giants of the microbial world (they're protozoans: parasites that, like bacteria, have a single cell, but which also have nuclei and some other cellular machinery), they are relatively easy to see and identify; and second, because it was a reliable killer of the most prolific lab animals, white mice. In 1903, Ehrlich had synthesized his own azo dye as a potential cure. Named trypan red in honor of its intended target, it followed a frustrating but common course: early success, followed by long-term failure. It didn't kill all versions of the protozoan, which made it functionally useless.

But not without providing useful data about the next step. In 1863, the French chemist (and one of Pasteur's many rivals) Antoine Béchamp

---

* Azo is an abbreviation of azobenzene, a blanket term for compounds made up of two rings comprising six carbon and five hydrogen atoms—phenyl rings—connected by two nitrogen atoms sharing a double bond of two electrons.
† Methylene blue, which is still used today for the treatment of urinary tract infections, would have a darker future. It was one of the drugs tested on human subjects at Auschwitz.

had discovered an arsenic-based compound he had named atoxyl; some recently published experiments indicated that atoxyl killed trypanosomes. In 1905, Ehrlich's by then well-cultivated talent for visualizing molecular structure resulted in a remarkable insight: The structure of atoxyl was not that of a chemically deactivated, or stable anilide, but of a far more changeable arsonic acid. The significance was huge: He could try to induce a chemical reaction from an anilide for years without getting any more response out of it than he would by hitting a steel girder with a rubber hammer. An arsonic acid, however, was chemically reactive. As his colleague Alfred Bertheim would later write, "Probably for the first time, therefore, a biologically effective substance existed whose structure was not only known precisely but also . . . was of a simple composition *and extraordinary reactivity, which permitted a wide variety of modifications*" [emphasis added]. In 1907, Ehrlich began tinkering with the molecule's promisingly unstable structure. He would continue tinkering for three years.

When he was asked, later in life, to describe his experimental strategy, Ehrlich responded with the ponderous "uniform direction of research combined with as much independence as possible for individual researchers." (It is even more ponderous in German: *Einheitliche Richtung der Forschung bei möglichst selbständigen Leistungen der Einzelnen.*) This sounds a bit more democratic than was actually the case; most of his subordinates recall a lot more uniformity of direction than independence; one of them remembered vividly that Ehrlich "often hammered on the anvil of his assistant's brains." However perceived at the time, it seems unarguable that among Ehrlich's many gifts as a scientist, his ability to organize the work of dozens of subordinates stood out. He had consciously adopted the same management style that had been pioneered by the synthetic dye industry, in which a prominent research director, Heinrich Caro, likened the process of testing new compounds to an "endless combination game [utilizing] scientific mass-labor."

The atoxyl experiments were among the very first in biology to apply these principles; to take a compound whose structure resembled other molecules that exhibited at least some of the desired effect (the term of art is "lead compound") and modify its chemical structure in a systematic and methodical way in order to optimize that effect. Though finding a promising compound like the organoarsenic atoxyl is good fortune, modifying it successfully depends on a conscious and deliberate strategy. Ehrlich's was driven by his belief in the side-chain receptor theory, which demonstrated that the action of any substance, helpful or harmful, was entirely a matter of chemical affinity: the right lock-and-key combination.

The goal was at the intersection of helpfulness and harm: Anything powerful enough to have an effect is virtually certain to have more than one. If it can kill a pathogen, it can kill (or at least damage) a host. The goal of the atoxyl experiments was to increase the damage the compound could inflict on the pathogen, while reducing its impact on the host. Ehrlich's hypothesis, based on long chemical experience with both medicines and aniline dyes, was that substituting different amine groups—the simple nitrogen-based structures that attached themselves to atoxyl's central ring like a kickstand on a bicycle—could lower toxicity to the host. One group of researchers in Ehrlich's lab busied themselves with finding just such an amine structure.

The other goal was increasing atoxyl's firepower against pathogens; a task made even more difficult because it wasn't clear where the firepower came from in the first place. Ehrlich postulated that what made atoxyl toxic to trypanosomes was not arsenic per se, but a particular arsenic *radical*: an atom with an unpaired electron in its outer shell. Ehrlich asked Bertheim to test this particular hypothesis by dumping electrons into, or reducing, atoxyl in the lab.* The one Ehrlich had his eye on was *trivalent* arsenic, one with three bonds and an electron in its outer shell. However, in a test tube, atoxyl was *pentavalent*, with five bonds to arsenic in its outer shell. It was also harmless. Somehow, the host organism itself was changing the struc-

---

* Technically, arsenic oxide. *Redox* (reduction-oxidation) reactions occur when atoms transfer electrons; reduction is the gain of an electron, oxidation the loss. Trivalent arsenic is reduced; pentavalent is oxidized.

ture of the arsenic, reducing the number of free electrons. This gave the lab's other team its task: hastening the reducing process that turned pentavalent arsenic into trivalent.

This sounds more rational than it actually was. The nice name for the approach is trial and error: brute force. Each of the new compounds, with increased amounts of trivalent arsenic, and different substitute compounds in the amine group, was tried out on test animals all over Europe.

Small wonder that the approach worked slowly. But it did work. It wasn't until arsenophenylglycine, known internally as Compound 418, was tested in 1909 that Ehrlich and his colleague, the Japanese bacteriologist Sahachirō Hata, found any real success, and it was decidedly mixed. Compound 418 cured sleeping sickness, but came with significant side effects, including an unfortunate habit of causing blindness in a small percentage of the animals on which it was tested.

On August 31, 1910, the lab hit the jackpot. Compound 606: arsphenamine.

Arsphenamine was not, as most histories have it, the 606th compound synthesized. Ehrlich's lab had been working hard, but not that hard. Hoechst's naming convention used the first digit to refer to a particular experimental compound, and the following numbers to denote a specific variant, which made arsphenamine the sixth version of the sixth compound. Because different compounds were being tested simultaneously, 606 had actually been synthesized for the first time in 1907, *before* number 418, which was the eighteenth version of the fourth compound.

Normally, variants that underperformed would have been discarded; and, indeed, Ehrlich's lab had moved on after the first tests of 606. The reason the lab returned to a compound discovered years earlier was the accidental discovery that arsphenamine didn't kill only trypanosomes. Because Ehrlich thought, mistakenly, that trypanosomes caused syphilis, he had tested the drug on a wide range of syphilitic animals as well, recruiting collaborators from all over the world, including Kitasato Shibasaburo, since returned to Tokyo, and Albert

Neisser, in what was then the Dutch East Indies, to test versions on rabbits, monkeys, and apes. Compound 606 cured them, too.

For good sound business reasons, this made arsphenamine far more interesting to a European chemical company. Sleeping sickness was a disease of what was not yet known as the Third World. Syphilis, on the other hand, had been killing and disabling Europeans for centuries. Though epidemiological historians continue to debate whether the disease was already present in Europe before explorers brought it back from the New World at the end of the fifteenth century, there's no doubt that it was one of the best-known and feared diseases in Europe from 1495 onward, causing everything from the characteristic genital sores, to painful abscesses, to destruction of the nervous system, to death. In 1520, the great humanist philosopher and poet Desiderius Erasmus called it ". . . the most destructive of all diseases," asking, "What contagion does thus invade the whole body, so much resist medical art, becomes inoculated so readily, and so cruelly tortures the patient?"

To ask the question was to answer it. Attempts to treat syphilis were either useless—the resin from the Caribbean "holy wood" known as guaiacum—or nearly as dangerous as the disease itself—mercury in its various forms, whose predictable side effects included mouth ulcers, loss of teeth, and even death, particularly since treatments could go on for years.* And, since blocking the route of transmission for the "great pox"—almost always sexual contact—was as problematic as asking people to stop breathing, syphilis was a terror for centuries before the organism that caused it was identified: the corkscrew-shaped bacterium *Treponema pallidum*.

A terror for some, an opportunity for others. With hundreds of thousands if not millions of men and women in the world's wealthiest

---

* From whence the phrase, "A night with Venus, a lifetime with mercury." It's not that mercury isn't a powerful bactericide. In fact, it is more than fifty times as deadly to bacteria as Lister's carbolic acid, which is why it was used, beginning in the 1930s, as a vaccine preservative in the form known as Merthiolate or thimerosal. Worth noting: In order to soothe public anxieties, and despite no evidence that thimerosal was harmful in the amounts used, it was removed from most vaccines beginning in 1999.

countries acquiring syphilis every year, a cure looked like a very profitable item for the company that could manufacture it in quantity.

By the first years of the twentieth century, the German chemical industry was a patchwork of partnerships and cooperative marketing arrangements—in German, *Interessengemeinschaft*, or "communities of interest." All of them were profitable, but like all profitable enterprises eager to become more so. One group was dominated by Agfa (*Aktiengesellschaft für Anilinfabrikation*, or the Aniline Manufacturing Corporation) and BASF (*Badische Anilin-und Soda-Fabrik*, or Baden Aniline and Soda Manufacturing). Another consisted of Hoechst AG—the same company where Behring had managed to exclude Ehrlich from participation in the profits from the diphtheria antitoxin—and Cassella Manufacturing.

The motto of Ehrlich's lab was *Geduld, Geschick, Glück, und Geld*— patience, skill, luck, money. Paying dozens of researchers to synthesize and test hundreds of potential molecules over the course of years isn't cheap. That's why the most significant innovation of the great German biochemical revolution was neither drugs nor vaccines themselves, but the creation of a durable funding source for ongoing lab research. Cassella had invested thousands of marks in Ehrlich's research—far more, even accounting for inflation, than Napoleon III had provided to Pasteur—and supplied dozens of different chemical compounds customized to Ehrlich's specifications, all in return for a share in any subsequent patents. This, even more than the drugs themselves, was new. It was also more than a little controversial, at least within the great chemical corporations, which had grown profitable on dyes and fertilizers rather than drugs. Ehrlich argued strenuously that it was also necessary, writing in 1908, "The material and mental support of our chemical factories is largely indispensable for modern therapeutics, and it would therefore not be advisable to loosen this natural union." In 1910, the "natural union" paid off. Cassella's partners at Hoechst AG introduced arsphenamine under the trade name Salvarsan. It was the world's first synthetic chemotherapeutic agent.

Salvarsan was a huge success, within a year of its introduction the most widely prescribed drug in the world. It was also one of the most challenging, for both doctors and patients. Common side effects included nausea and vomiting. Storage of the compound was a tricky business, requiring vials tightly sealed to avoid oxidation. Treatment entailed weekly injections of the highly diluted compound over the course of a year—*very* highly diluted; each injection required at least 600 cc of solution per injection, or a pint and a quarter of solvent pumped into a patient's body with every visit to the doctor. Many doctors decided to experiment with different solvents; others to try different injection locations: into the muscle, under the skin, or directly into the bloodstream. Intramuscular and subcutaneous injections tended to be painful; intravenous ones still unfamiliar to physicians, and therefore risky. The dangers of such a regimen, given the needles and syringes available in the first decade of the twentieth century, were very real: Too much seepage from the injection site could result in amputation or even death. Moreover, as would become a familiar theme of every subsequent advance in drug therapy, it was widely used for ailments well outside its arena of effectiveness. In some unfortunate cases, it resulted in death from cerebral hemorrhage. In 1912, Hoechst, Ehrlich, and Hata responded with a new and improved version, marketed as Neosalvarsan, which was soluble in water, and less toxic to boot. Like its precursor, it was a blockbuster: a miracle drug.

Even a century after its introduction, the structure of Salvarsan remained unknown; that is, though the formula for Salvarsan is well known, Ehrlich's lab never did establish the precise structure of the compound. They originally synthesized it by a very delicate chemical process* that tended to introduce impurities that could affect the compound's toxicity and/or efficacy dramatically. The actual structure of Salvarsan, which wasn't discovered until 2005, consists of mixtures of cyclic molecules, which prompted some people to suggest that it wasn't a *Zauberkugel*, but *Zauberschrot*: a magic buckshot. Even more mysteriously: As of this writing, no one has yet figured

---

* Reducing phenylarsonic acid with dithionite. Because dithionite is an anion, or charged compound, of sulfur and oxygen, it was very difficult to work with.

out the mechanism by which Salvarsan and Neosalvarsan target *T. pallidum* so precisely.

The precision of the arsphenamines is one reason that, while Ehrlich's achievement looms as large as any in the history of medicine, it was also a false start. Though it would remain the most important drug in the medical arsenal for decades, it was too narrowly effective to revolutionize the practice of medicine all by itself. The few attempts that followed in its wake were notably less successful; Ehrlich himself developed another compound intended to cure streptococcal pneumonia. However, while optochin (also known as ethylhydrocupreine hydrochloride) is indeed toxic to the strep bacteria, it is nearly as dangerous to humans, in whom it frequently causes irreversible blindness. Today it is used, like Koch's tuberculin, as a diagnostic tool to identify the presence of *Streptococcus pneumoniae* rather than a therapy.

Nonetheless, Salvarsan marked as huge a step on the way to the birth of the antibiotic age as the germ theory itself. The power of a well-financed mass attack on disease had been demonstrated. It might very likely have provided the template for a true therapeutic revolution even before Ehrlich's death in 1915, had Europe not chosen that very moment to destroy itself in the most violent war in the history of Western civilization.

Within a week of the assassination of the heir to the throne of the Austro-Hungarian Empire in Sarajevo, France and Germany had declared war on one another, troops of the Dual Monarchy were shelling Belgrade, Tsar Alexander had mobilized the armies of the Russian Empire, Kaiser Wilhelm's army had invaded Belgium, and Great Britain had declared war on Germany. Europeans would spend the next four years killing, maiming, poisoning, and starving one another.

This grim future was unknown to the Russians, French, Germans, Austrians, and Britons who applauded the start of hostilities, each convinced that a rapid and painless victory was theirs for the taking.

Two months later, the cheering hadn't quite stopped, but it probably should have. During the first eight weeks of combat on what came to be known as the western front, the armies of the German Empire had suffered 550,000 casualties; those of the Republic of France 590,000. At the first Battle of Ypres alone, which bloodied the fields of Flanders for five gory weeks, from the nineteenth of October to the twenty-second of November 1914, 85,000 French and 56,000 British soldiers were either killed or wounded, virtually destroying the relatively small British Expeditionary Force. On the other side, more than 18,000 German soldiers were listed killed or missing, and more than 29,000 wounded.

One of them was a nineteen-year-old onetime medical student from the University of Kiel named Gerhard Domagk. At the beginning of hostilities, he had enlisted in a Leibgrenadier Regiment from Frankfurt-von-Oder that was, literally, decimated at Ypres: more than one soldier in ten killed or wounded. With nothing salvageable of his unit, he was transferred to the eastern front and the German army's medical department, named, with apparently no irony, the "Sanitary Service." As might be expected, the field hospital to which Domagk was assigned in the Ukraine lacked something on the sanitary front. It also lacked any real weapon against the infectious diseases that killed as many soldiers during the First World War as gunfire: cholera, typhus, gangrene, dysentery, and a hundred more. Domagk would spend the rest of the war seeing medical impotence close up, treating patients and assisting in surgeries. By the end of 1918, and the Armistice, he had had more than enough of war, but not of medicine. Three years later, he graduated from Kiel as a fully accredited physician, and went to work in the Baltic port city's main hospital.

He recognized quickly that he was far better suited to a career as a researcher than as a clinician, and joined the Pathological Institute at the University of Greifswald, as a lecturer, a privatdozent, in 1924. There, under the institute's director, the pathologist Walter Gross, whom he would follow to the University of Münster, he began his studies of the most powerful weapon (really, the only weapon) against infectious disease: the vertebrate immune system. In a series of experiments, he injected hundreds of mice with a known pathogen, the bacterium known as *Staphylococcus aureus*, which causes everything from skin infections to pneumonia, and then extracted cells from the lining of the animals' livers known to be part of the innate immune system* to see how many of the staph bacteria they had gobbled up.

Credit: Wellcome Library, London

Gerhard Domagk, 1895–1964

* These cells are known as "Kupffer cells," for the German physician, Karl Wilhelm von Kupffer, who first observed (and misidentified) them in 1876. They are part of the family of white blood cells that patrol for, and destroy, invading foreign substances.

Domagk's experiments resulted in two significant findings. First: The performance of the Kupffer cells improved with exposure to the staph pathogen (though he couldn't know the reason, which wasn't discovered until the 1980s: pattern recognition proteins known as Toll-like receptors that identify different sorts of toxins produced by bacteria and tailor the response by white blood cells sent to destroy them). The second revelation was the big one: A staph cell that had been weakened beforehand by exposure to antiseptics was easier for the Kupffer cells to destroy. The insight would determine the next thirty years of Domagk's life. In 1927, he followed his boss Walter Gross once again to a job in the same industry that had served as patron to Ehrlich and Behring: the dye business.*

Two years before, the German chemical and dye business had made a giant leap in the consolidation it had been pursuing before and especially after World War I. As far back as 1904, Carl Duisberg, the managing director of Bayer, had been lobbying his competitors on the value of combination, in the form of a fifty-eight-page memo laying out the advantages: lower costs, shared patents, lower risk, and *much* higher profits. He persisted through the First World War (during which Bayer's U.S. subsidiary, known primarily for its Bayer Aspirin brand, was confiscated as enemy property) and the hyperinflation of the 1920s, in 1925 finally calling a conference—never one to shy away from grandiosity, he named it the "Council of the Gods"—at which the eight largest chemical firms in Germany, including Agfa, BASF, and Hoechst/Cassella, merged into one, to be known as the "Community of Interest of Dye Businesses"—in German *Interessengemeinschaft Farbenindustrie*, or I. G. Farben. Carl Bosch of BASF was its chief executive, Duisberg its chairman of the board.

I. G. Farben was the largest chemical company in the world, and one of the largest of any sort, comparable in size to American companies like General Motors or U.S. Steel, with interests not only in dyes,

---

* Gross would, in his way, become as notable as Domagk, though for a far blacker achievement. A raging anti-Semite and early member of the Nazi Party, he created and ran the party's Office of Racial Policy from 1933 until his suicide in 1945.

but in photographic film, industrial solvents, and, with the acquisition of BASF, fertilizers. Bosch had developed a method for synthesizing ammonia—the Haber-Bosch process, the discovery of which would produce two separate Nobel Prizes in Chemistry, one for Fritz Haber in 1918, the other for Carl Bosch in 1931—that today produces 150 million metric tons of fertilizer annually and is responsible for feeding nearly half of the world's population. Most relevantly for Domagk, I. G. Farben was also Germany's largest manufacturer of pharmaceuticals, such as they were.

They weren't much. Though Behring's antiserums and Ehrlich's arsenicals were widely used (and highly profitable), and Bayer's antiprotozoal drugs Plasmoquine and mepacrine (also known as Atabrine) were starting to be used as malaria treatments, the pharmacopoeia available to fight infectious disease wasn't significantly greater than it had been when George Washington awakened on the last morning of his life. Nonetheless, the possibility of antibacterial drugs was still promising enough that Heinrich Hörlein, head of pharmaceutical research at Bayer, set Domagk up in a newly refurbished research lab that Carl Duisberg had built in Elberfeld, a Westphalian town just east of Düsseldorf.

Duisberg and Hörlein believed they knew why no successful antibacterial drugs had appeared since Ehrlich's Neosalvarsan a decade previously. It was the same argument behind the formation of I. G. Farben itself: scale. If Ehrlich had tested dozens of different recipes in order to find the antisyphilis treatment, Bayer would try hundreds. Or thousands. The way to make trial and error work was simply to increase the number of trials to the point that an effective compound would be guaranteed to emerge: to do for chemical innovation what Henry Ford had done for automobile manufacture.*

And so they did. Beginning with Domagk's arrival in 1927, two

---

* By the end of the twentieth century, investigators working at what came to be known as "combinatorial chemistry" realized the limits of such an approach, which is that there are far more ways to combine atoms than there are atoms in the universe. As a result, even if some imaginary supercomputer had started testing a new combination every second since the Big Bang some fourteen billion years ago, it would barely have scratched the surface of all possible molecules.

chemists from Bayer's tropical medicine group, Josef Klarer and Fritz Mietzsch, started producing coal tar–based compounds with at least some antibacterial properties on the test-tube-and-ring stand equivalent of the assembly line at Ford's River Rouge factory. By 1931, they had delivered more than three thousand such compounds to Domagk's lab. There, the bacteriologist exposed them, one by one, to the family of *Streptococci* bacteria, pathogens responsible for diseases ranging from the familiar and moderate—strep throat, impetigo—to deadly diseases like toxic shock, streptococcal pneumonia, meningitis, and such exotica as the flesh-eating disease known as necrotizing fasciitis. Most of the compounds supplied for testing against strep were created by modifying chemicals that, as with Ehrlich's early researches, were originally dyes that showed some affinity for a particular bacterium by staining it.

Meanwhile, Domagk was working at an equally feverish clip to isolate a strain of *Streptococci* that was a consistent and reliable killer of laboratory mice. His purpose was to create a kind of ideal experimental bacterium, which could rapidly show the effectiveness—or ineffectiveness—of each of the compounds supplied by Klarer and Mietzsch. Over the first few years, his supercharged strep produced thousands of rodent corpses, each one accompanied by autopsy notes documenting the particular compound used to treat it, the progress and symptoms of the disease(s) at time of death, and the method by which the bacterium and the compound had been exposed to the unlucky rodent. The notes also, of course, had a place for showing which compound had a significant effect against the strep bacteria.

For years, the last space remained stubbornly blank. A decade later, Iago Galdston, a physician and the secretary of the Medical Information Bureau of the New York Academy of Medicine, wrote, "By 1930 it was the universal opinion of physicians that nothing could be discovered which would be effective against the ordinary diseases produced by bacteria."

Less than a year later, in 1931, Mietzsch and Klarer started modifying azo dyes, believing them to be less toxic than aniline dyes, and therefore more promising. By summer, the promise appeared to be borne out. One of the azo-derived compounds—KL-487, for "Klarer,

#487"—killed at least some of Domagk's strep strain, without over-whelmingly toxic side effects. Others followed: KL-517, KL-529. The protocol employed looks, in retrospect, a lot like fairly arbitrary tinkering. The first attempts involved successively adding side chains composed of chlorine atoms, one at a time, to the azo base. When the possibilities of chlorine were exhausted, they moved on to arsenic, then to iodine.

Eventually, on the advice of Hörlein himself, Klarer tried sulfur as the missing ingredient in an azo compound. The first one he tried, KL-695, used another important chemical included in the dye business known as sulfanilamide—formally para-amino-benzene-sulfonamide—that had been around since 1909. In late 1932, Domagk treated some more of his unfortunate mice with a compound that integrated sulfanilamide with an azo dye.

After four years of failure, it's not hard to imagine the exultation when the results of the new compound were revealed. Domagk had administered the sulfanilamide-plus-azo compound to twelve mice infected with his superstrep strain, with fourteen untreated mice as a control. Within a week, all fourteen untreated mice were dead, most within two days. All twelve that received the compound survived. Whether administered intravenously or orally, it completely cured strep infections in mice. By the time Klarer and Mietzsch had delivered KL-730 to Domagk, the conclusion was inescapable. The lab had found the world's first successful antibacterial drug.

Or, actually, drugs. Virtually all azo dyes with a sulfanilamide side chain worked on strep infections, which meant that patenting all conceivable variations would be a legal nightmare for Bayer. Unlike tangible goods, which can be sold only once, intellectual property can be sold to multiple consumers without diminishing inventory, a horn of plenty that never empties.* This means that patents are valuable to inventors because they alone gain the legal right to bar anyone else from profiting off the invention. In the early days of patents, legal

---

* The economist Kenneth Romer calls it *nonrivalrous* property, as distinguished from *rivalrous*.

protection stopped at national borders, making patents far more valuable to inventors working in a large country like France or Britain
than in a relatively small one like the Netherlands or Switzerland.
Patents also behave very differently on mechanical inventions than
on chemicals. No one can sell a new engine while keeping its components secret from a competitor, but it is considerably harder to reverse
engineer a dye or a drug. One consequence is that, although patent
offices first started granting licenses in the seventeenth century, the
Swiss, the Dutch, and, before the 1870s, the Germans, who specialized in chemical innovation, tended to avoid even seeking them. So
long as secrecy could be preserved while dyes and other chemicals
were being sold, patents offered far more risk than reward.

As chemistry matured into a science that could decipher the secrets
of drugs and dyes, techniques for analyzing the components and structures of new chemical compounds made it relatively easy to reproduce
a competitor's proprietary molecule. And once secrecy offered no protection to novel chemicals, patents became as necessary for them as
they had long been for machinery. Because of their different histories
with patent protection, the Swiss, Dutch, Germans, and even the
French still guarded chemical innovation with systems very different
from the British and American model. German chemical patents didn't
protect a novel *product* like a new chemical compound, but rather the
process by which it was synthesized. Bayer didn't need to patent every
effective variant of the sulfanilamide-plus-azo compounds, but they
did have to publish a detailed description of the method by which all of
them were created, which was itself a risky proposition. There are
multiple methods of chemical synthesis, and a competitor could, if
clever enough, create a sufficiently different one, and so benefit from
Bayer's discovery while investing only a fraction of Bayer's time and
money. So when Bayer applied for a new patent on Domagk's discovery in 1932, the application was quite deliberately as obscure as possible
about the creation of KL-730 in order to protect its exclusivity.

For the next two years, the new drug, which had been named
Streptozon, was tested on both animal and human subjects. By the time
the patent was granted, in 1934, Domagk had demonstrated that the

drug worked against a number of nonstrep infections as well, including spinal meningitis, some strains of *pneumococci*, and gonorrhea. As a result, Bayer changed the name of the new drug. They called it Prontosil.*

In 1935, Domagk finally announced his discovery in an article for the journal *Deutsche Medizinische Wochenschrift*, and the number of human trials expanded to include subjects in Great Britain. There, the first results were considerably less encouraging than the ones reported from the German tests, probably because in the United Kingdom it was tested on strains of *streptococci* different from the supercharged version developed by Domagk. Notably, however, one of them was *S. pyogenes*, the primary cause of what was then known as puerperal, or childbed, fever.

The oldest known medical texts record the dangers, and frequency, of fevers suffered by women within hours of giving birth. The risks of contracting such a fever, however, skyrocketed with the growth of hospital births during the nineteenth century. By the time the Hungarian physician Ignaz Semmelweis, then working at Vienna General Hospital, published a study of childbed fever in 1847, it was attacking as many as four new mothers in ten. Tellingly, Semmelweis discovered that the risks of childbed fever were significantly lower in home births than obstetrics wards. The cause, in the days before Lister, was the way physicians practiced their craft: never washing their hands.† Improved hygiene and antisepsis reduced the numbers of victims substantially, but did not eliminate the risk of disease. Though Semmelweis and others had made puerperal fever less common, their technique was prevention, not cure. Throughout the 1920s, physicians tried Salvarsan and other arsenicals, failing miserably. As late as 1936, it was still

---

* The new name was actually an old one: The dye at the center of KL-730 had originally been developed as a rapid-working dye for leather, when it was called "Prontosil Rubrum" or "fast red."

† In 1843, Dr. Oliver Wendell Holmes, Sr.—the same one who sneered at the entire nineteenth-century pharmacopoeia—quoted another doctor as saying, "I had rather that those I esteemed the most should be delivered unaided, in a stable, by the mangerside, than that they should receive the best help, in the fairest apartment, but exposed to the vapors of this pitiless disease. Gossiping friends, wet-nurses, monthly nurses, the practitioner himself, these are the channels by which . . . the infection is principally conveyed." Holmes was quoting James Blundell's *Lecture on Midwifery*.

attacking up to three hundred out of every ten thousand new mothers in the United States . . . and killing forty-nine of them. Until, that is, the discovery that Prontosil stopped *S. pyogenes* even more effectively than Ehrlich's magic bullets targeted *T. pallidum*. Virtually overnight, mortality from childbed fever fell from 20 to 30 percent to 4.7 percent.

By then, the power of Prontosil had already been demonstrated to the Bayer scientists in the most personal way possible. In December 1935, a needle was driven into the hand of Gerhard Domagk's six-year-old daughter, Hildegarde. Lasting injury from the trauma was unlikely, but her risk of infection considerable, as the following days proved. An abscess formed, and Hildegarde's temperature sky-rocketed, peaking at more than 104°. *Streptococci* had entered the girl's bloodstream. A year earlier, she would very probably have lost her arm, and possibly her life. But this was the year of Prontosil. A week after starting the therapy, her infection was beaten.

The effectiveness of Prontosil was no longer in doubt. The mechanism by which it performed its magic, however, remained a mystery. Did the sulfa chain activate the azo dye, or the other way around? Why did Prontosil cure *streptococcal* infections in infected animals, but fail to kill strep bacteria in a test tube? Throughout Europe and the United States, biochemists tried to solve the puzzle, including Leonard Colebrook at St. Mary's Hospital in London, and Ernest Fourneau, head of pharmaceutical research at the Institut Pasteur in Paris, both of whom immediately requested samples of Prontosil for testing. Colebrook's request was granted. Fourneau's was not.

Fourneau's history with Bayer generally, and Hörlein particularly, was less than collegial. Despite Pasteur's early experience conducting research on behalf of the French wine and cheese industries, France had scrupulously avoided strong links between commerce and chemistry. The premier research facility in the country, the Institut Pasteur, had originally been funded by donations from French families with no expectation of a return on their investment. As a result, the institute, employing a few dozen researchers at most, was unable to produce anything like the number of new chemical compounds that were created by hundreds of Bayer and I. G. Farben scientists and technicians. Nonetheless, Bayer regarded Fourneau as a formidable competitor. For one thing, he was a superb chemist; the author of more than two hundred scientific papers, Fourneau had already developed a synthetic alternative to cocaine for his then-employer, Camille Poulenc of Établissement Poulenc Frères,* before joining the Institut Pasteur as its director of research in 1911. For another, he had the French patent system on his side. Since Bayer drugs and dyes had no legal protection outside the borders of Germany, they were fair game for anyone with the energy to decode even the most opaque patent. Fourneau was nothing if not energetic; he devoured German academic journals and patent filings, attended German scientific meetings and trade fairs, and worked long hours building a file showing every niche where a homegrown drug might replace an import. In 1921, he had permanently alienated Duisberg and Hörlein when he reverse engineered Bayer's proprietary treatment for sleeping sickness, originally introduced under the brand name Germanin, but sold in France by Poulenc as Stovarsol.

Scarcely surprising, therefore, that when Hörlein received Fourneau's request for samples of Prontosil, he stalled. Fourneau did not. He assembled a team of French chemists to copy the existing drug using the obscure language of Bayer's patent, and by the middle of 1935, Rubiazol, the French version of Prontosil, was on the market.

To Bayer, this was an expected, though exasperating, cost of doing business. What followed, however, was startling—and disastrous. On November 6, 1935, Daniel Bovet, a member of Fourneau's team engaged in testing Rubiazol, infected forty mice with *streptococci*. The population was then broken into ten groups of four mice each: one an untreated control group, another with the original version of Prontosil/Rubiazol, and seven more with new products just synthesized at the Institut Pasteur. Bovet would later write, "I only had seven new products and we had an extra group of four mice. Why, I asked, not just try the product common to all these products?" The common

* In 1928, Poulenc would merge with Société Chimique des Usines du Rhône— roughly, Rhône Chemical Factory Inc.—to form Rhône-Poulenc.

product was the side chain that Klarer and Mietzsch had added to their azo dye: sulfanilamide.

Within days, the results were tallied. The control group of untreated mice had all died, as had six of the seven groups treated with the newly synthesized compounds. The four mice injected with Prontosil/Rubiazol were alive. So were the four given pure sulfanilamide. "From that moment on," wrote Bovet, "the German chemists' patents had no more value."

And they knew it. At virtually the same moment that Bovet was testing his mice in Paris, in Elberfeld, Klarer and Mietzsch tested a compound—KL-821—that was nothing but sulfanilamide alone: no azo dye. It not only worked just as well as Prontosil, it was even more useful, treating staph infections as well as strep. The reason that Prontosil worked only in vivo and not in vitro—in a living animal and not a test tube—was finally clear: Living animals produced enzymes that separated the dye from sulfanilamide.

It was disappointing enough that all those experiments with different azo combinations had turned out poorly. The really bad news was this: Since a Viennese chemist named Paul Gelmo had identified sulfanilamide as part of his 1908 doctoral thesis and patented it in 1909, the substance was now in the public domain.

At Bayer, the news was devastating. No one yet knows how so many skilled researchers could fail so totally—the original documents remain sealed away—but a good guess is that Prontosil was a particularly acute example of the cognitive bias that psychologists call "functional fixedness" and civilians know as "if the only tool you have is a hammer, everything looks like a nail." The large chemical industries had been built on dyes and wore blinders that shielded them from just about anything else. Which also explains why, even after Bayer knew that adding azo dyes to sulfanilamide did literally nothing to improve the drug's antibacterial effectiveness, and that the active ingredient in the drug was freely available to anyone with the 1935 equivalent of a home chemistry set, they *still persisted with the launch of Prontosil.*

They tried to brand their product, so far as they could, selling something they called Prontosil Album (white Prontosil, also known as pure sulfa) alongside Prontosil classic for a year. It helped, a bit. Sales were strong and getting stronger when, in November 1936, the son of the president of the United States, Franklin Delano Roosevelt, Jr., contracted a vicious strep infection and, after weeks at death's door, was miraculously cured by the still-experimental drug. The front-page headline for the December 17, 1936, edition of the *New York Times* read:

YOUNG ROOSEVELT SAVED BY NEW DRUG

DOCTOR USED PRONTYLIN ON

STREPTOCOCCUS INFECTION OF THE THROAT

CONDITION ONCE SERIOUS

But Youth, in Boston Hospital, Gains Steadily;

Fiancée, Reassured, Leaves Bedside

This wasn't all bad for Bayer. "Prontylin" was the trade name under which sulfanilamide was sold in the United States by the Winthrop Chemical Company, which was half owned by Bayer's parent, I. G. Far- ben. It was even manufactured at the old Bayer factory in Rennselaer, New York, which had been seized during the First World War. But the combination of unprecedented American demand and a complete lack of patent protection had the predictable result. By the end of 1937, a hun- dred different companies were selling sulfanilamide under a hundred different names. And not just in the United States; a Japanese version was named Pratanol; a Dutch one, Streptopan. In Brazil it was being sold as Stopton, in Czechoslovakia as Supron. The French had five different versions; the British, more than thirty. Forty-six years later, the physi- cian Lewis Thomas wrote, "I remember the astonishment when the first cases of pneumococcal and streptococcal septicemia were treated in Bos- ton in 1937. . . . Here were moribund patients, who would surely have died without treatment, improving in their appearance within a matter of hours of being given the medicine and feeling entirely well within the next day or so. . . . Medicine was off and running."

Off and running, though not always in a helpful direction. For the first time, though far from the last, a truly effective medicine was wildly overprescribed. Physicians used it for head colds. Expectant mothers were regularly given sulfa prophylactically, before they showed any signs of puerperal fever.

The immediate consequences of the sulfa craze were significant enough. The side effects of the drug, including nausea and painful crystals in the urinary tract, affected hundreds of thousands of patients who got no benefit in return. The first high-profile sulfa treatments are just as historically consequential, though, marking the moment in time when large numbers of patients first demanded a specific therapy from their physicians by name. It would be some time before it was accurate to refer to the sick as "consumers" of health care, but that's how they—and what were not yet known as the "worried well," patients with no need for doctoring, but a lot of demand for it—started to behave. Physicians, finally armed with a treatment that actually cured infectious disease, were only too happy to oblige.

Sequelae is the word physicians use to describe the chronic and persistent consequences that follow an acute episode: back pain after an automobile accident, for example. The long-term sequelae of the public embrace of sulfa include the belief that doctors ought to be able to cure just about anything. One result was that everything else they provided—typically comforting their patients with knowledge of the likely course of a disease or condition—started to be devalued by both doctors and patients. Antibiotics, and the transformation they ignited, professionalized health care in a way like nothing before or since. That professionalism was a huge benefit, but it wasn't cost free.

Nor was sulfa itself. Among other things, the drug remained difficult to administer, since it wasn't easily soluble in water, or really much of anything else. This limited its appeal to patients, particularly in the United States, who preferred to take their medicine orally. This problem, all by itself, led to the single largest scandal in the history of antibacterial drugs, as well as the birth of drug regulation in America.

When the sulfa craze began in 1937, laws protecting patients from the dangers of the new drugs weren't unknown, but they might as well have been. Though the first code of ethics of the American Medical Association barred direct-to-consumer advertising of drugs from "ethical medical practice," the first federal law to address the administration of powerful and dangerous compounds in service of medicine was the Biologics Control Act of 1902. It was followed by the Pure Food and Drugs Act of 1906, which set out penalties for adulterating or misbranding medicines. Those penalties were determined by the United States Department of Agriculture's Bureau of Chemistry, headed from 1906 to 1913 by Harvey Washington Wiley, whose feelings about pharmaceutical regulation can be summed up by the title of a book he wrote in 1929: *The History of a Crime Against the Food Law: The Amazing Story of the National Food and Drugs Law, Intended to Protect the Health of the People, Perverted to Protect Adulteration of Food and Drugs.*

Even when Wiley wrote his "amazing story," most of the drug business remained in patent medicines, a predictable consequence of a widespread American belief in self-medication (or "autotherapy"). The number of advertised compounds grew from about a thousand in 1858, generating about $3.5 million in annual sales, to twenty-eight thousand by 1905, with revenues of nearly $75 million. By 1912, that number had increased to more than $110 million. To call them fraudulent is to compliment them. Some of the most popular were targeted at the same market that would eventually turn Viagra into a multibillion-dollar business: "Persenico" promised to combat "low vitality . . . of sexual origin" while "Revivio" simply told men they could "improve your vigor." Those that weren't ineffective were frequently poisonous; "Gouraud's Oriental Face Cream" contained possibly toxic levels of mercury. The most famous of all patent medicines wasn't quite that dangerous. The recipe that Confederate army veteran John Pemberton concocted to wean himself from an addiction to the morphine he had been given after the Battle of Columbus, a mixture of alcohol, coca leaves, and kola nuts that was first sold as Pemberton's French Wine Coca, and later, sans alcohol, as Coca-Cola, took the nation by storm beginning in 1886.

The standardization of "real medicines dates back as far as 1820, when one of Benjamin Rush's former students, Jacob Bigelow, published the first edition of a catalog of more than two hundred drugs he entitled the United States Pharmacopeia." Nonetheless, it took nearly a century before the 1906 Pure Food and Drugs Act established the USP as a national formulary standard and used it to require that drugs not be adulterated. It also set out punishments for misbranding, such as selling under a false name, or not identifying the presence of narcotics, like morphine, opium, cocaine, and heroin. The word "penalties" should be used advisedly, however, since they only called for confiscation of adulterated or misbranded drugs, not prosecution of dealers.* Moreover, because the original law focused exclusively on labeling, rather than safety, it was virtually useless. Robert N. Harper's "Cuforhedake Brane-Fude" was marketed as an entirely safe "brain tonic" containing alcohol, caffeine . . . and acetanilide, a highly toxic analgesic that causes cyanosis. When the government, under Harvey Wiley, won the maximum penalty of $700, it represented a tiny fraction of the thousands the compound had earned.

And so it went. In 1911, the 1906 act was amended to allow a misbranding prosecution if the package included statements that were "false and fraudulent," which created a loophole big enough to drive an entire industry through. Remedies that promised to cure everything from pneumonia to tuberculosis to cancer were protected so long as the manufacturer could claim that it *believed* the claims, and thus had no fraudulent intent.

Which was the state of play when Franklin Delano Roosevelt, Jr., became the poster boy for the new wonder drug, sulfa. Though his father's New Dealers (primarily Walter G. Campbell, one of Wiley's protégés, and the agricultural economist Rexford Tugwell) had been trying to expand the reach of the Pure Food and Drugs Act ever since they took office in 1933, they failed each and every time.

They probably would have gone on failing, if sulfa had been an easily soluble powder.

---

* In fact, in 1911 the U.S. Supreme Court decided in *United States v. Johnson* (221 US 488) that a concoction sold as a cancer remedy wasn't, by the terms of the act, misleading, since it didn't misrepresent the "strength, quality, or purity" of the compound. Misleading therapeutic claims weren't the government's business, since any claim of therapeutic efficacy, in those days, was protected opinion.

The S. E. Massengill Company of Bristol, Tennessee, was one of the dozens of American patent drug manufacturers eager to get on board the sulfanilamide gravy train. The sales staff at Massengill, and Samuel Evans Massengill himself, believed that the most successful version of the new drug would resemble cough syrup: a sweetened liquid. Harold Watkins, the company's chief chemist, tried a number of solvents before he happened on a solution of 58 pounds of sulfa, along with raspberry flavoring and saccharine, dissolved in 60 gallons of diethylene glycol, a component of resins, brake fluid, and coolants whose properties in living animals—dizziness, intoxication, and nausea an hour after ingestion; within days, an elevated heart rate, muscle spasms, and acute kidney failure—were evidently unknown to him. In October 1937, Massengill's Elixir Sulfanilamide went on sale.

On October 11, 1937, the president of the Tulsa County Medical Society sent a telegram to the American Medical Association's chemical lab, alerting them to six deaths that occurred shortly after administration of the elixir. Eight days later, the *Washington Post*'s headline read "VENEREAL DISEASE 'CURE' KILLS 8 OUT OF 10 PATIENTS IN OKLAHOMA." On October 25, the *New York Times* ran a story about the "nationwide race with death."

The response from Washington was uncharacteristically decisive. The Food and Drug Administration put its entire field force—239 inspectors—to work tracking down every drop of Massengill's initial shipment: 240 gallons of the diethylene glycol–laced sulfa syrup. They secured 234 gallons and 1 pint. It is presumed that most of the remaining 5⅞ gallons never made it down the throats of anyone, since fewer than a hundred people were killed.*

Massengill, despite protesting its innocence (and blaming sulfa, rather than diethylene glycol, for the deaths), was brought to trial, though not, as one might expect, for poisoning its customers. In doing so, it hadn't actually broken any law. Instead, it was prosecuted

---

* The actual provable count was seventy-three, but no one seriously believes that all were accounted for. One additional victim was Harold Watkins, who committed suicide while awaiting trial.

under the only permissible statute: for mislabeling, since they had marketed the deadly syrup as an "elixir" and elixirs were required to contain alcohol. In fact, the existing law only allowed FDA agents to seize bottles that had been shipped across state lines, and whose seals had not been broken. Massengill eventually pleaded guilty to 174 counts of mislabeling, fined $150 each plus costs, for a total of $26,100.

Nonetheless, the scandal had at least one salutary effect: The 1938 Federal Food, Drug, and Cosmetic Act was signed into law by President Roosevelt on June 25, 1938. It was, in many ways, a dramatic improvement over existing law. Section 502, for example, extended the "misbranding" violation to include "adequate warnings for use in those pathological conditions or by children where its use may be dangerous to health, or against unsafe dosage or methods or duration of administration . . ." and required listing of all ingredients, not simply the "active" ones (diethylene glycol, after all, hadn't been the active ingredient in the Massengill syrup). Secret remedies were now prohibited. Interstate shipment was forbidden without approval from the secretary of agriculture. The act's final clause created new regulations for a new category: "New Drugs." Henceforth, manufacturers needed governmental permission to market any drug not already available.

It's hard to know whether in 1938 anyone knew just how many "new drugs" would appear over the next ten, twenty, and fifty years. For the immediate future, however, sulfa remained king. In 1939, a variant known as sulfapyridine, developed in Great Britain by the firm May & Baker, a subsidiary of Rhône-Poulenc, was used to quell a meningitis outbreak in Sudan—significantly, one caused by *meningococcus*, not *streptococci*. Sulfapyridine was also far more effective against *pneumococcal pneumonia* than its predecessors like Prontosil, reducing mortality in one test from 27 percent to 8 percent. Popularly known as M&B 693, it was approved for sale by prescription in the United States in March 1939 (and would famously cure Winston Churchill from a strep infection in December 1943).

Also in 1939, Gerhard Domagk was awarded the Nobel Prize in Physiology or Medicine. He was forbidden to appear in Stockholm to accept the award by his government, then only months from plunging

Europe into a war more impactful than the First World War of 1914–1918, but offended at the Peace Prize given to the German pacifist Carl von Ossietzky in 1936 for exposing German rearmament. Domagk was even briefly jailed for being "too polite to the Swedes" in his response to the prize.*

By then, the British chemists Donald Woods and Paul Fildes had discovered the mechanism by which sulfa drugs do their magic. Sulfa inhibits an enzyme essential for the production of the B vitamin required for folate production in bacteria. Sulfanilamide essentially tricks bacterial enzymes into latching onto it, rather than the correct compound, para-aminobenzoic acid, or PABA (which is why sulfa remained limited in its effectiveness; not all bacteria use PABA).

Despite those limits, sulfa was used throughout the Second World War to treat everything from wound sepsis to gonorrhea. The first wonder drug, however, was no more immune to the resourcefulness of bacteria than any of its successors.

The phenomenon known as acquired antibiotic resistance is no simple thing, and is imperfectly understood even today.† Bacteria are exquisitely sensitive to environmental change; the single-celled organisms are quick studies of evolutionary innovation, able to acquire new genetic blueprints for just about every imaginable function from other bacteria, and even from free-floating viruses. Moreover, they reproduce so quickly—some bacteria have generations only twenty minutes long—that a useful bit of DNA spreads extremely rapidly. Under strong selection pressure, such as the presence of a powerful bactericide like the sulfanilamides, a new gene that codes for a folate-producing enzyme unaffected by sulfa appears seemingly overnight. Which is why, given its widespread use in wartime, the phenomenon

---

* In 1947, he finally collected his medal, though not, by Nobel rules, his cash award. In his Nobel Lecture, on "Further Progress in Chemotherapy of Bacterial Infections," he observed: "The problem . . . could be solved neither by the experimental medical research worker nor by the chemist alone, but only by the two together working in very close cooperation over many years."

† Antibiotic resistance refers specifically to pathogenic bacteria. However, the same phenomenon can, and does, occur in fungi, viruses, and multicellular parasites.

of sulfa resistance appeared quickly. The first soldiers treated for gonorrhea with sulfa experienced a 90 percent cure rate in the late 1930s. By 1942, the rate was 75 percent and falling fast.

By then, however, Domagk's miracle was about to be preempted by a new weapon against bacterial disease; not just a single magic bullet, but an entire arsenal.

THREE

"Play with Microbes"

The consulting room of Dr. Colenso Ridgeon, KBE, was, in the spring of 1903, located on the second floor of an unremarkable building in London's Marylebone neighborhood, indistinguishable from a hundred other Victorian apartments. In addition to an overstuffed couch and chairs, the green-wallpapered room featured a marble-topped console table with gilt legs ending in the claws of a sphinx; a mirror so covered with paintings of palms, lilies, and other plants that it was no longer useful as a reflecting surface; and a writing table, on which a microscope, test tubes, and a small alcohol stove fought for space with piles of papers and journals. Like the doctor himself, the room made some concession to modernity—it had replaced gas lighting with electricity—but otherwise looked, in the first years of the twentieth century, almost exactly as it had in the middle of the nineteenth: a very model of bourgeois solidity.

If Dr. Ridgeon's place of work was conservative, the work he did there was positively revolutionary. So revolutionary, in fact, that in June 1903 he received a knighthood for it. After thirty years as a practicing physician, Dr. Ridgeon had discovered a cure for tuberculosis, an historic achievement if anything was.

You will, however, search in vain through any history of medicine looking for him, or his discovery. Colenso Ridgeon was and remains a creature of the imagination whose genius (and taste in furniture) was described, and his knighthood awarded, in the first act of a 1906 play entitled *The Doctor's Dilemma*, written by the Irish iconoclast George Bernard Shaw.

Shaw was a Fabian Socialist, a sometime atheist, an antivivisectionist, and an Irish patriot, but his real passion was satire, and he had written the play to skewer the medical establishment of his day. Debating the "dilemma"—whether Ridgeon should use his tuberculosis cure to treat an honorable but inconsequential friend, or a deeply immoral but brilliant artist, a choice complicated by the doctor's lust for the painter's wife—are a group of physicians with what might charitably be called blinkered approaches to the healing profession. A onetime schoolmate of Dr. Ridgeon, Dr. Leo Schutzmacher, offers the two-word secret to his successful practice—"Cure Guaranteed"—while confiding, "You see, most people get well all right if they are careful and you give them a little sensible advice." Cutler Walpole is a frighteningly eager surgeon who believes all disease is a version of blood poisoning, which is invariably cured by the removal of an entirely invented organ he calls the "nuciform sac." Sir Ralph Bloomfield Bonington, having read a page or two from Koch and Pasteur, treats each of his patients with the most dangerous microbes he can find, expecting thereby to promote a natural cure. "Drugs are a delusion. Find the germ of the disease; prepare from it a suitable antitoxin; inject it three times a day quarter of an hour before meals; and what is the result? The phagocytes* are stimulated; they devour the disease; and the patient recovers." Of them all, only the now-retired Sir Patrick Cullen escapes Shaw's barbs, pointing out to Ridgeon that, despite his newfound cure, "I've known over thirty men that found out how to cure consumption. Why do people go on dying of it?"

For a satire to work at all, its targets must be familiar to its audience, and so it was with *The Doctor's Dilemma*. Every one of the London playgoers who attended its premiere would have known physicians who reflexively removed each of their patient's tonsils or appendix irrespective of their presenting symptoms; or those who monomaniacally prescribed antitoxins or antiserums for everything from skin lesions to cancer. And they would have recognized the

---

* *Phagocyte* was a then-current term for the white blood cells known today as macrophages, discovered in 1884 by the Russian biologist Ilya Mechnikov.

model that Shaw used for Colenso Ridgeon: the nation's most famous physician, Dr. Almroth Edward Wright.

In 1906, Wright, a friend and occasional debating opponent of Shaw, was forty-five years old, and already frequently referred to in the press as "Britain's Pasteur." He had come to medicine by a somewhat circuitous route; before graduating as a physician from Trinity College, Dublin, in 1883—Wright was partly Irish by descent—he had already taken First-Class Honours in modern literature, and would later read "Jurisprudence and International Law with a View to the Bar." From 1895 on, though, he was completely devoted to medicine, and his particular area of interest was the discovery of an immunization against typhoid fever, one of the deadliest diseases in history.

Variously blamed for the fifth century B.C.E. Plague of Athens, the destruction of the English colony of Jamestown in the seventeenth century, and the deaths of tens of thousands of American Civil War soldiers, typhoid fever is one of a dozen different diseases caused by a strain of bacteria from the genus *Salmonella*, specifically *S. typhi*.* Typhoid fever is a killer, but not a sudden one; a typical infection lasts up to a month, beginning with a characteristic low-grade fever and slowed heartbeat. As the fever rises, the heart continues to slow, frequently accompanied by delirium. The bacteria reproduce rapidly, causing the host's abdomen to distend and severe diarrhea to follow. Internal organs like the spleen and liver enlarge. Intestines hemorrhage, and sometimes perforate, spreading infection to the internal membrane: the peritoneum. Unless the victim's immune system is successful in fighting off these sequential attacks, death results; mortality can run as high as 25 percent in untreated typhoid fever.

The destructiveness of any infectious disease is a function of both virulence and mobility: how much damage the pathogen causes, and how easily it travels. Because *S. typhi* spreads by consumption of

---

* In a moderately perverse bit of medical history, the family of bacteria that causes typhoid fever—a disease that affects only humans—is named for Daniel Elmer Salmon, a doctor of veterinary medicine, whose assistant first isolated the bacterium in 1885, and named it for his boss.

drinking water contaminated with human feces and urine, it was particularly deadly wherever large numbers of people with poor access to
sanitation lived in close contact: the poorer quarters of rapidly growing nineteenth-century cities, for example. Or—even more lethally—
armies in the field. During the Spanish-American War, typhoid fever
killed more American soldiers than either battle wounds or even the
feared viral disease known as yellow fever.

For obvious reasons, then, military doctors were particularly concerned about the disease, none more so than Wright, whose first job
after becoming a physician was at the Army Medical School at Net-
ley, near Southampton. There, in 1895, he developed the first effective
vaccine against typhoid, inoculating subjects not with a weakened
version of the *S. typhi* bacterium (this was the method tested by Robert Koch), but, far more safely and effectively, a dead one. He tested it
first on himself, then on fifteen volunteers, and finally on a regiment's worth of British soldiers headed to India. It was Wright's first,
and greatest, triumph: Of nearly three thousand subjects, only ten
contracted the disease.

Despite this success, he was unable to persuade the conservative
policymakers in Britain's War Department to inoculate troops being
sent to South Africa in 1899 to fight in the Second Boer War. As a
result of their reactionary hostility to even modern medicine, even in
the face of what seems its inarguable success in India, over the next
three years typhoid proved more deadly to the British army than the
combined efforts of the Afrikaner Transvaal Republic and the Or-
ange Free State. At least twenty thousand British soldiers contracted
the disease, and more than nine thousand of them died. Disgusted,
Wright left the War Department in 1902, and moved to St. Mary's
Hospital. On Praed Street in London's Paddington neighborhood, St.
Mary's was one of the last of London's so-called "voluntary" hospitals, set up for the care of the working poor, and one of the first
conceived of as a teaching hospital with an attached medical school.
There, following the model of Pasteur and Koch, he opened a
laboratory—the "Inoculation Department"—that he would direct for
the next forty-five years.

Credit: Wellcome Library, London

Almroth Wright, 1861–1947

In retrospect, Wright's accomplishments at St. Mary's never soared as high as his reputation, and his place in history has suffered in consequence. During his lifetime he was hugely famous and influential, eccentric and intimidating. In legend, at least, his memory was so remarkable that he had committed a quarter of a million lines of poetry to it. Wright was tall and striking, careless of his dress but wickedly entertaining in his speech, a brilliant raconteur and lecturer, and a public figure whose opinion was sought on issues scientific, social, and political until his death in 1947. He was also a fine and innovative experimentalist, and a master of laboratory technique. With little more than a microscope, a Bunsen burner, and a supply of rubber nipples and glass tubing, he was able to perform extraordinarily sophisticated research* in ways that remind historians of science just how much the

---

* In 1912, he wrote a classic book on laboratory methods entitled *Technique of the*

lab once depended on the steady hand of a craftsman. The vials that Wright used to collect blood, his so-called blood capsules, were custom-made bits of glass pipette that he melted and drew in the flame of a Bunsen burner into narrow tubes that he then bent at an angle. Snipping the glass at one end provided a needle, and the curve allowed the glass straw to draw the blood by capillary action.

Wright's lab skills were, however, something of a two-edged sword, since they made him prone to accept his acute clinical observations as irrefutable proof. Yet he was hopeless with numbers; the original dead-cell typhoid inoculation he performed on 2,835 India-bound soldiers was almost certainly successful, but you couldn't prove it by Wright's statistics. According to Britain's leading mathematical biologist, Karl Pearson, the data Wright collected were useless for concluding anything: no control groups, no attempts to show what statisticians call the "null hypothesis"—the assumption that there is no relationship between two phenomena, such as "being inoculated" and "getting typhoid." Wright's statistical illiteracy was likely a consequence less of his temperament than of his eccentric education, which was almost willfully deficient in practical mathematics, even by the standards of nineteenth-century Great Britain. Wright had been home tutored, and spent far more time on Latin declensions and the history of the common law than on regression analysis.* It is almost certainly that blind spot that explains his devotion to one of the great dead ends in medical history: vaccine therapy, the use of substances that activate the adaptive immune system to fight a specific disease as a therapy, rather than a preventative.

Wright was a vaccine absolutist, famously observing, "The physician of the future will be an immunizator." The key, to Wright, was the particular character of an individual patient's immune system, *not* an attack on pathogens using chemicals such as Ehrlich's Salvarsan or, later, Domagk's sulfanilamide. This debate—whether disease was best

---

Teat and Capillary Glass Tube, a title that is responsible for a large but unknown number of cases of uncontrollable giggles among generations of medical students.

* It wasn't that the required mathematical tools hadn't yet been invented. The German mathematician Carl Friedrich Gauss had taught the world how to minimize measurement errors in the first decade of the nineteenth century.

understood as what occurs when a healthy host encounters a pathogen, or as the consequence of what happens when an unhealthy host does so, with the latter providing evidence of a deficiency in the host's internal environment—dates back to Pasteur, and, in some senses, remains alive today.* Convinced by the promising results of serum therapy to treat illnesses such as rabies, Wright predicted that similar techniques could be used "to exploit the uninfected tissues in favor of the infected." Wright named this phenomenon the "opsonic mechanism."

In explaining the wholly fictional tuberculosis cure at the heart of *The Doctor's Dilemma*, Colenso Ridgeon says to Sir Patrick Cullen that "opsonin is what you butter the disease germs with to make your white blood corpuscles eat them." Shaw was prescient. A story in the *New York Times* from March 31, 1907—a year *after* the premiere of the play—is headlined: "THE NEW HOPE FOR TUBERCULOSIS: DISCOVERY OF 'OPSONINS' PROMISES TO REVOLUTIONIZE MEDICINE." The newspaper goes on to quote Wright on his discovery that opsonins do their work "by uniting with the micro-organisms, the invading germs, and rendering them more palatable, so to speak, to the white corpuscles."[†]

Opsonins are real. Any molecule that enhances the way white blood cells ingest and kill invading pathogens is, technically, an opsonin, as are those that activate the complement that is part of the innate immune system. Opsonic therapy, however, never lived up to its initial promise; nor did Almroth Wright. "Britain's Pasteur" almost certainly saved hundreds of thousands of lives during the First World War; the British army that fought on the western front was given Wright's typhoid inoculation, and only twelve hundred soldiers died

---

* The original conflict about the nature of disease, between Pasteur and his friend and fellow physiologist Claude Bernard, has been co-opted by twenty-first-century advocates of alternative medicine; some use it to attack the germ theory itself.
† The *Times* continued: "The opsonic theory explains for the first time the undoubted value of blisters and poultices and the old-time counter-irritants. Take the case of an open wound. If the blood is poor in opsonins, the wound will refuse to heal, bacteria will attack its lacerated surfaces, and suppuration will set in. . . . The poultice encouraged the flow of blood and lymph to the infected area."

of it, out of more than two million. He performed heroically during the war itself, demonstrating the limits of antiseptic pastes and liquids like Lister's carbolic acid to treat battlefield wounds—carbolic acid didn't just attack pathogens, but the immune system's leukocytes as well—and the dangers of airtight bandages, which encouraged the growth of nasty, gangrene-causing bacteria like *Clostridium perfringens* that thrive in anaerobic environments. Nonetheless, he is now mostly remembered as Britain's leading opponent of women's suffrage. And, of course, as the inspiration for Colenso Ridgeon, Shaw's dilemma-facing doctor.

This scants Wright's real legacy: the Inoculation Department he founded and ran for decades at St. Mary's, and that he made into an incubator for the next generation of antibacterial researchers. One of his subordinates there, who followed him to France, was Leonard Colebrook.

Another was a Scottish physician named Alexander Fleming.

When Fleming joined Almroth Wright at the Inoculation Department in 1906, the then twenty-five-year-old physician was a promising if not yet accomplished researcher. He was also a sort of anti-Wright—where Wright was tall and physically imposing, Fleming was short and slender; Wright had a mustache that made walruses envious, Fleming was clean-shaven; and while Wright was never happier than when speaking publicly, Fleming was so self-effacing that students had to strain to hear his lectures. He had graduated with distinction both from the Royal Polytechnic Institution (now the University of Westminster) and from St. Mary's Hospital Medical School, where he had trained as a surgeon before discovering a talent for experimental research. In 1909, Fleming designed a new test for syphilis that required less blood, and was more effective, than the eponymous diagnostic invented three years before by the German bacteriologist August Paul von Wassermann. The following year, Fleming began working with Leonard Colebrook to investigate the properties of Ehrlich's magic bullets: Salvarsan and Neosalvarsan.

When the First World War broke out, Fleming and Colebrook accompanied Wright to France, where the core of St. Mary's Inoculation Department joined the British military hospital at Boulogne-sur-Mer. That hospital, one of several built to accommodate the huge number of casualties from the first Battle of Ypres—the same battle in which Gerhard Domagk received his wound—would prove a remarkably productive research facility, even without considering the circumstances under which it had been established. While working at the hospital, the St. Mary's team discovered that the standard of care for wounds—antiseptic ointments and airtight bandages—actually promoted infections rather than preventing them. As Fleming recognized, the cause was the variety of bacteria that grow even in the absence of oxygen,* particularly under the skin's surface. And those bacteria were, literally, everywhere, even after exposure to powerful antiseptics. It took some experimental skill to demonstrate why.

As he described in a now-classic paper written for the *Lancet*, Fleming exposed two sets of glass tubes to a highly concentrated bacterial soup. One set was left whole, while the other was broken to create a ragged edge that would simulate a battlefield wound. After both were washed with antiseptics, the unbroken test tubes were completely disinfected, but the bacteria in the broken tube's hidden recesses stubbornly reappeared, even after washing in carbolic acid. Fleming had demonstrated experimentally why even unbloodied uniforms from soldiers with supposedly disinfected wounds remained rife with pathogens. Dangerous ones. Fifteen percent of battlefield wounds contained staph, 30 percent tetanus, 40 percent strep . . . and 90 percent were infected with the gangrene-causing *C. perfringens*.

In November 1918, the First World War ended. For Fleming, now returned to St. Mary's, the war against pathogenic bacteria was just

---

* A huge number of dangerous infections are caused by such *anaerobic* bacteria, not just gangrene, but tetanus and peritonitis. Anaerobes come in two basic flavors: *obligate* anaerobes like *clostridia*, for which oxygen is a poison; and *facultative* anaerobes, including *staphylococci* and *streptococci*, which are able to grow without oxygen, but can often use it if it is present.

getting started. He had acquired a more sophisticated understanding of the resourcefulness of his opponents, but was no closer to victory over them until, in 1922, he made his first improbably accidental discovery. As his laboratory assistant, V. D. Allison, later recalled, Fleming:

> . . . *was busy one evening cleaning up several Petri dishes which had been lying on the bench for perhaps ten days or a fortnight. As he took up one of the dishes in his hand, he looked at it for a long time, showed it to me, and said: "This is interesting." . . . It was covered with large yellow colonies which appeared to me to be obvious contaminants. But the remarkable fact was that there was a wide area in which there were no organisms. . . . Fleming explained that this particular dish was one to which he had added a little of his own nasal mucus, when he had happened to have a cold. The mucus was in the middle of the zone containing no colony. The idea at once occurred to him that there must be something in the mucus that dissolved or killed the microbes. . . .*

Fleming named the substance found in his mucus *lysozyme*: the first purely organic substance shown to have antibacterial properties. However, the unlikelihood of the discovery as reported seems almost too much to credit. First, Fleming later revealed that the mucus had *accidentally* dripped from his nose onto one of the Petri dishes. Not just dripped, but dripped onto the one Petri dish that had, somehow, picked up a bacterial contaminant from a fortuitously open window . . . and, even less probably, since most bacteria (and all important pathogens) are unaffected by lysozyme, the contaminant on the twice-lucky dish would have had to be one of the few bacteria with lysozyme sensitivity. This is the laboratory equivalent of buying winning lottery tickets twice on the same day.

However improbable its discovery, lysozyme was an interesting, but relatively inconsequential compound, one that Fleming accurately recognized as an enzyme: a large molecule that increases the speed of organic chemical reactions. Some years later, it was identified as one of the components of the body's innate immune system, whose activity works to damage bacterial cell walls. This is a nontrivial ability that offers some protection against infection, particularly in newborn children, but isn't much use against most pathogens. The same can't be said of Fleming's next encounter with good fortune, which occurred some five years later.

Alexander Fleming, 1881–1955

Credit: Wellcome Library, London

The canonical story of the discovery of penicillin is eerily similar to the one describing the chance discovery of lysozyme. As Fleming later recalled, he had sloppily left Petri dishes containing staph cultures unattended on a bench in his St. Mary's lab when he departed for vacation in August 1928. When he returned, on September 3, he found that one of the Petri dishes had been contaminated, again via a conveniently open window, this time by a fungus. The evidence was even more startling than five years earlier: Around the fungal contamination was a ring in which all the *staphylococci* had disappeared. Something had killed them.

For weeks, Fleming worked to cultivate the fungus, *Penicillium notatum*, technically a mold (molds—in Britain, "moulds"—are fungi that take the form of tiny multicellular filaments, which gives them their characteristically fuzzy appearance; unicellular fungi are yeasts). The mold was producing some substance that was deadly to the staph bacteria, a substance that Fleming first named "penicillin" in March 1929 in an article entitled "On the Antibacterial Action of Cultures of a Penicillium, with Special Reference to Their Use in the Isolation of *B. Influenzae.*"

So: another open window. Another accidental contamination. Another brilliant discovery illustrating the power of serendipity.

Or, perhaps not. For decades, historians and scientists have puzzled over the inconsistencies in Fleming's account. For one thing, the famous window in Fleming's lab (today preserved as a museum) appears to have rarely, if ever, been opened in order to avoid exactly what Fleming later claimed: the accidental introduction of a contaminant. For another, the timing seems as fuzzy as the fungus itself: The original account has the Petri dishes left alone for more than five weeks, but in 1944 Fleming himself said that the effect was observed after only two. Though Fleming's chronology puts the discovery of the world-historic Petri dish on September 3, the first day in which he noted it in his lab notebooks is October 30.

Most tellingly of all: If the *Penicillium* mold had appeared after the staph colonies were well established, they would have killed it long before it could produce penicillin in the first place. Fleming could not, of course, have known this, but the way penicillin kills or degrades *staphylococci* is by disrupting the mechanism the bacterium uses to build new walls during the process of cell division—which means that it works only when the bacteria are dividing.* Unless the mold is present *before* the staph, no ring of bacterial death.

This doesn't mean that Fleming was fraudulent, or even forgetful, either in his account of the discovery of lysozyme or of penicillin. The most appealing explanation for the discrepancies between either account and, well, logic, is something else: playfulness.

Fleming grew up in a family that was notoriously fond of cards, table tennis, and quizzes. As an adult, he was an avid player of croquet and snooker, and a skilled rifle shot (he was originally recruited to St. Mary's as a shooter for the hospital's famously competitive sports teams). As a golfer, he far preferred what might be called creative versions of the game, up to and including using his putter as a pool cue. He painted, too; not well, but certainly inventively: When he was a young researcher at St. Mary's, Fleming regularly created images—Madonna and Child; the logo of St. Mary's; the Union Jack— on Petri dishes, using agar as the canvas and microorganisms that turned different colors as they grew as pigments. (It's worth noting that this kind of "painting" demanded an almost terrifying level of both bacteriological knowledge—which ones turn red, which green, and when—and hand-eye coordination.) One of his biographers observed: "Fleming's natural level was indeed play. . . . 'I play with microbes'—his often repeated description of his work—was literally true. Most of his research was a game to him and indeed most of his enjoyment came from games of all kinds."

But while Fleming loved to play with microbes, he was almost completely at sea when it came to human interaction: a terrible conversationalist and a worse lecturer. He was painfully shy, and had no interest in discussing either his methods or results, which is why it's at least plausible that Fleming's penchant for games, combined with his natural reticence, persuaded him to cast the lysozyme discovery as if it had been the equivalent of drawing a winning hand at bridge.

Understanding why he would trot out such a similar story for the

---

* More about this below.

far more important discovery of penicillin requires some additional context. The first, and most important, fact about the discovery is that hardly anything about it was documented at the time. Six months would pass before Fleming published his results in 1929, and penicillin would remain at best a novelty, at worst a dead end, for another decade. By the time the magnitude of the discovery came to light, it's certainly possible that the details had faded in Fleming's memory, or that he recalled a sequence of events reminiscent of his earlier discovery without any intention to deceive. The debate about his reasoning (if any) and motives (ditto) continues, with no resolution in sight.

This wouldn't be very important except for the significant matter of credit, which, as we shall see, would become a very thorny issue indeed. Insofar as any one person is associated with the discovery of penicillin—of antibiotics generally—in the public consciousness, it's likely to be Alexander Fleming. It was considerably easier for Fleming to become the first hero of the antibiotic age if his great discovery were understood to be one he himself recognized immediately, rather than one about which he was mistaken—which was apparently the case, at least for a few months. The story that better fits the facts, one that was persuasively argued by the physiologist Robert Scott Root-Bernstein, is that Fleming hadn't been doing a staph experiment at all; instead of testing a number of different pathogens, he was actually observing a number of different fungi. He was, therefore, surprised by the accidental introduction of the staph colonies, and not the other way around.

Which leaves unanswered the experiment's goal, about which Fleming was famously unhelpful. Experiments need hypotheses, after all. If his goal was to observe staph colonies under different conditions, what were they? If, instead, his objective was something different, the problem vanishes. Root-Bernstein's answer? Fleming was seeking a new source for lysozyme.

In 1928, lysozyme was still Fleming's only notable discovery; even after he had become world famous, it remained so important to him that he felt obliged to observe, in his Nobel Lecture in 1945, that "Penicillin was not the first antibiotic I happened to discover. In 1922,

I described lysozyme, a powerful antibacterial ferment which had a most extraordinary lytic [i.e., destructive] effect on some bacteria." And molds were an extremely promising source for antibacterial compounds like lysozyme. As early as 1876, John Tyndall described the effects of the mold: "On the 13th [December 13, 1875] a thick blanket of *Penicillium*" that had formed on meat "assumed a light brown colour, as if by a faint admixture of clay . . . the slime of dead or dormant *Bacteria*, the cause of their quiescence being the blanket of *Penicillium*. I found no active life in this tube, while all the others swarmed with *Bacteria*." Joseph Lister himself had noted that bacteria grown in samples that included molds—formally, the one known as *P. glaucum*—failed to grow. It is suggestive that Fleming's original notes show that he thought the original bacteria-inhibiting fungus was *P. rubrum*, not, as the mycologist Charles Thorn later showed, *P. notatum*. This is a lot easier to explain if Fleming was testing a number of different molds, in an ongoing search for something he *knew* existed—lysozyme—rather than something never before seen. He had been working diligently since 1922 examining mucus, sputum, blood, plasma, eggs, snail slime, flowers, and even root vegetables, looking for lysozyme, and the best explanation for the inconsistencies in Fleming's years-later account is that he initially believed that he had found, in the *Penicillium* mold, not a new antibacterial compound, but an old one.

However dubious his later recollections of the discovery, Fleming deserves enormous credit for recognizing the potential of the compound he first called "mould juice" and, even more, for the experiments that followed. He and his assistant, Stuart Craddock, became penicillin farmers, cultivating crop after crop of *Penicillium*, and fertilizing them with protein from the heart of a bull. With sufficient quantities to begin testing, they established where the juice was effective—it killed not just *staphylococci*, but also *streptococci*, and a number of other bacteria—and where not—typhus, for example, was immune to it. And they established that it was harmless to nonbacterial cells. Though it would be more than thirty years before the mechanism was understood, the reasons for penicillin's effectiveness against some, but not all, bacteria (and its lack of toxicity toward animal cells) were the same.

Since 1884, biologists had known that some varieties of bacteria stained a distinctive color when exposed to certain dyes, usually gentian violet. Such bacteria were named Gram-positive, for Hans Christian Gram, the Danish biologist who had discovered the phenomenon; those that didn't take the stain were, predictably enough, Gram-negative.

When bacteria reproduce, they behave like a water balloon with a string tied around the middle: The original cell divides in half, while the cell wall is stretched and twisted. Penicillin (and related compounds) chemically weakens the cell walls of the bacteria as it splits—this is why penicillin only acts on bacteria when they're dividing—but only for Gram-positive bacteria that are unprotected by the lipopolysaccharide outer membrane: staph and strep, but not typhus.* And not animal cells, which have membranes but not walls, and are therefore safe from the actions of penicillin.

Fleming and Craddock could demonstrate these facts empirically without knowing their underlying mechanisms (Fleming, in particular, was convinced that the compound was, like lysozyme, an enzyme). They could purify the "mould juice" and distill it, though not to the point that they were able to remove an unpredictable number of toxic impurities, nor did their fluids ever reach a concentration of more than 1 percent. They could even test it on a sinus infection afflicting Craddock. Unaccountably, though, Fleming never tried it on an infected animal. In the words of another researcher, "All Fleming had to do to demonstrate the curative effect of penicillin was to inject .5 ml of his culture fluid in to a 20 g mouse infected with a few streptococci or pneumococci. . . . He did not perform this obvious experiment for the simple reason that he did not think of it."

One possible reason for the failure to test the fluid on a test animal

---

* The Gram-positive/Gram-negative distinction isn't as clean as introductory biology textbooks suggest. Some taxonomists recognize as many as twelve different bacterial phyla, only two of which are "totally" Gram-positive.

was the difficulty of working with it; nothing Fleming and Craddock did could solve the problem of penicillin's instability: Within days, sometimes hours, the stuff, evaporated now into syrup, would lose its effectiveness. As a result, though Fleming would return to penicillin research occasionally over the following three years, he actually spent more time during the 1930s on lysozyme. As late as 1940 he wrote of penicillin, "The trouble of making it" even as a local antiseptic "seemed not worthwhile." He had run up against the wall separating bacteriology from biochemistry, and St. Mary's had no one with the training needed to hurdle it. Craddock later recalled that "we knew very little when we began" using the classic technique of dissolving the compound in a solution of acetone (sometimes ether) and allowing it to evaporate. "We knew just a little more when we had finished."*

St. Mary's deficiency in well-educated chemists was both absolute and relative, especially as compared to Germany. Because German bacteriological research was largely taking place in industrial chemistry laboratories, the level of expertise in processes that occurred at the molecular level was almost unimaginably high. St. Mary's Inoculation Department was probably Britain's most sophisticated and successful bacteriological research facility; in May 1932, the Dean of St. Mary's, Sir Charles Wilson (later Lord Moran) appealed to England's greatest newspaper mogul, Lord Beaverbrook, asking for the £100,000 needed to build St. Mary's into a world-class institution. Beaverbrook did come through with a generous contribution, around £60,000, but that was probably less than Klarer and Mietzsch spent just producing test compounds for Gerhard Domagk. Competing with I. G. Farben with nothing but donations from Britain's aristocrats was a bit like a prep school taking the field against the New York Yankees.

The German dye conglomerates were even able to transform physicians into adequate chemists; in the case of Paul Ehrlich, brilliant

---

* Even experienced chemists found penicillin a puzzle. Harold Raistrick, a biochemist at the University of London, was equally unable to purify the mercurial—in this case, only metaphorically—compound.

ones. It was no accident that the argument at the core of Ehrlich's 1908 Nobel Lecture supposed that the future of microbiology would be chemical rather than observational. For decades few outside Germany listened.

Or, if they did, they did not understand. Almroth Wright, despite his very impressive skills, was poorly equipped to appreciate the importance of chemical analysis and synthesis. Worse, Wright's conviction that any successful attack on bacterial pathogens would emerge from the immune system rendered him uninterested in hiring competent chemists. In consequence, when Domagk's sulfa drugs ignited a revolution in medicine in 1934, Fleming's 1929 paper "On the Antibacterial Action of Cultures of a Penicillium" was all that existed to record his most important contribution to humanity's war on infectious disease.

At his death in March 1902, Cecil John Rhodes, the English-born founder of the British South Africa Company, was one of the wealthiest and most celebrated men in the world. He had built a hugely successful business empire—among other things, he founded De Beers Consolidated Mines, then and now the world's largest diamond mining concern—alongside one of the more traditional sort: the eponymous colony, later independent country, he christened Rhodesia. However, since Rhodesia was renamed Zimbabwe* in 1980 by an administration that found his brand of colonialism more than a little offensive, the Rhodes name lives on today only in institutions founded after his death: Rhodes University in South Africa, and, of course, the Rhodes Scholarships.

The scholarship, which funds two or three years of study at one of the University of Oxford's residential colleges, is awarded annually to college graduates from current and former British colonies, the United States, and—when world wars don't intervene—Germany. As of this

---

* Technically, Southern Rhodesia. The northern half of the colony became the country of Zambia in 1964.

writing, there have been 7,600 Rhodes Scholars, including dozens of men and women who are even more celebrated than their benefactor: Nobel Prize winners, generals, a few professional athletes, cabinet members, senators, governors, Canadian and Australian prime ministers, and even one U.S. president. It's a close call which of them has had the greatest impact on history, but there's little doubt about the first Rhodes Scholar to change the world, a physician who departed Australia for the journey that brought him to Magdalen College in January 1922. His name was Howard Florey.

Florey was then twenty-three years old, the youngest of five children born in Adelaide to a first-generation Australian mother and an English émigré father. He had excelled at both St. Peter's Collegiate School and at the University of Adelaide in every academic subject except math,* and at an exhausting number of athletic pursuits, including tennis, cricket, and football. His father's death in 1918, while Florey was studying for his medical degree, freed him to apply for, and accept, the scholarship that brought him and sixty-one others to Oxford,† a freshly minted MD ready to begin his studies at the university's Department of Pathology.

Pathology—broadly speaking, the study of the causes of disease—was, for obvious reasons, a young discipline in 1922, no older than the discoveries of Koch and Pasteur fifty years before. Oxford had started teaching it as a course in 1894, and achieved departmental status only in 1901. The same year Florey arrived in Oxford, the pathology department received a gift in the amount of £100,000 from the trustees of the estate of Sir William Dunn, a Scottish merchant banker and politician who had, like Cecil Rhodes himself, made a fortune in nineteenth-century South Africa. Though the planning and construction of the Sir William Dunn School of Pathology took four

---

* Florey would later recall that his inability to master higher mathematics was the reason he turned to medicine rather than physics; unlike Almroth Wright, however, he was mathematically competent enough to know what he didn't know . . . and to seek out help when it was needed.

† One of them, John Fulton from Missouri, will reappear in our story.

years—it wouldn't open its doors until 1927—Florey's timing could scarcely have been better.

In 1923, the young Australian earned First-Class Honours from the School of Physiology and the Francis Gotch Medal awarded to the department's most promising researcher. In his notebook of that year, the secretary of the Rhodes Trust, Sir Francis Wylie, described Florey as "a first-rate man—ranks with our best." Wylie wasn't the only one to recognize his talent. One of Florey's instructors, Sir Charles Sherrington, nominated him for the John Lucas Walker Studentship at Cambridge's Gonville and Caius College in 1923, which paid a stipend of £300 a year—around £15,000, or $24,000, today—plus another £200 for equipment.

Florey's second year in England was productive. He published four papers on a variety of topics, served as medical officer for an Arctic expedition, and, despite what even Florey himself recognized as a less than appealing personality—in 1923, he described himself in a letter to his future wife, Ethel Reed, as "developing into a rather nasty product"—even acquired friends; one of them, Charles Sutherland Elton, would become one of the founders of modern population ecology. He had found a mentor in Sherrington, who wasn't just professor of physiology at Oxford, but, in 1923, president of the Royal Society, the world's first and most distinguished scientific association, which meant that his endorsement was about as good as it got for an ambitious young researcher.*

During his third year he made his first connection with the Rockefeller Foundation.

Established in 1913 by the Rockefeller family "to promote the well-being of humanity throughout the world," in 1923 the foundation was the world's largest philanthropic enterprise, with a special interest in medicine and health. It had already established the world's first School of Public Health at Johns Hopkins University, and a dozen

---

* Sherrington, a brilliant and innovative scientist himself, would share the 1932 Nobel Prize in Physiology or Medicine with Cambridge's Edgar Adrian for "their discoveries regarding the function of neurons."

more public health universities around the world, working to eradicate parasitical diseases like malaria and yellow fever.

There were a number of reasons behind the decision of the world's wealthiest family to embark on a crusade to give away millions of dollars. One was surely to repair some serious image problems. After decades of building the Standard Oil trusts, and accusations of brutal competitive tactics in order to corner the entire oil industry, John D. Rockefeller, Sr., was not the most admired man in America. Well-meaning (and well-publicized) philanthropy could reverse some of this, but even there, the path was fraught. The behavioral and social sciences, including anything that touched on the nature of industrial relations, were strictly forbidden, especially after the 1914 Ludlow strike—better known as the Ludlow Massacre—in which dozens of miners and their families were killed at the Colorado mine owned by John D. Rockefeller, Jr. Pure science was the ticket: the purer, the better.

The really novel aspect of the foundation, though, was how the money was disbursed: Rockefeller, who had neither the time nor the inclination to decide on the foundation's grantees (and in any case his involvement was widely regarded as the next thing to poisonous), wanted experts doing the selection. Scientists would recommend other scientists, the best and the brightest that could be found. One of the foundation's executives, Wickliffe Rose, spent nearly a year in Europe asking the continent's most prominent researchers for nominees that might deserve grants from the foundation's International Education Board in order to—his words—"make the peaks higher" and let the results flow down the mountain to everyone else.* So-called circuit riders of the Rockefeller Foundation traveled the world, checkbook in hand, looking to identify the "future leaders in science," and both Cambridge and Oxford were regular stops on their journeys.

In 1925, Howard Florey received his first Rockefeller Foundation

---

* Between 1923, when it was founded, and 1938, when it was folded back into the Rockefeller Foundation, the IEB made grants to fifty-seven institutions and 603 Fellows, including Niels Bohr and Enrico Fermi.

fellowship, spending nine months, from September 1925 to May 1926, at various Rockefeller-funded labs in New York, Chicago, and Philadelphia, where he would establish a relationship with Alfred Newton Richards, the future president of the National Academy of Sciences.

Florey's transatlantic connections would prove enormously important, as, too, would his links to continental Europe—he had spent part of 1922 and 1923 at labs in Copenhagen and Vienna—though forging them proved a challenge. By the 1920s, scientific research had become so much more international than it had been when Koch and Pasteur were dueling over the relative importance of vaccination and sanitation as to be unrecognizable. German, French, British, Swedish, and American scientists collaborated regularly, published together, and—this isn't a contradiction—still fought fiercely over discoveries and priority. National affiliation wasn't completely forgotten, but the real competition was between laboratories, not nations.

All that collaboration and competition, however, wasn't free. Over the course of the next two decades, two different schemes for the funding and direction of scientific research were on offer. The industrial model, which had produced Paul Ehrlich's Salvarsan, and would, soon enough, be responsible for Gerhard Domagk's sulfa drugs, had the advantage of tremendous focus and discipline, but it was also highly dependent on confidentiality. The philanthropic model, on the other hand, benefited from collaboration and the ability to follow multiple lines of investigation, but even with the world's wealthiest families supporting it, it was still underfinanced. Worse, unlike corporations like I. G. Farben, philanthropists, whether individuals like Sir William Dunn or trusts like the Carnegie and Rockefeller foundations, lacked ruthlessness. They were temperamentally ill-suited to disinvest from programs rapidly, and, as a corollary, foundation-led research had difficulty reinforcing success. Even with Lord Beaverbrook's money, Alexander Fleming's department at St. Mary's couldn't afford to hire the chemical talent needed to exploit the original penicillin discovery. The same story—brilliant and promising research in a constant scuffle for money—was about to play out at Oxford, with a somewhat different conclusion.

In 1926, Florey married Ethel, his former medical school class-mate, now a doctor herself. Even by the standards Florey had estab-lished for bluntness and interpersonal tone deafness in his professional life, his courtship, conducted via slow-motion correspondence be-tween England and Australia, had been stormy. It was calmness itself compared to the marriage that followed, though, which featured score settling of a particularly vitriolic (and difficult to read) nature. By the time the Floreys' marriage was five years old, Ethel was complaining that her husband had sabotaged her career, while he, in turn, accused her of desertion, lack of affection, a disappointing fre-quency and variety in their sex lives, lousy cooking, and poor per-sonal hygiene, even reminding his—deaf—wife that she was "not a physically normal woman."

As miserable as Florey was in his marriage—or, to be accurate, as miserable as both Floreys were—it did little to divert his career. In 1927, he received his PhD from Gonville and Caius College at Cam-bridge, where he was appointed a lecturer in special pathology—a discipline about which he had known nothing before Sherrington took an interest in him, but which would be the subject to which he would devote the rest of his life.

Establishing himself in the field of pathology meant studying pathogens. For two years, Florey produced a huge volume of work, on subjects as varied as cerebral circulation, capillary action, and mucous secretions: an "experiment a day, including Sundays," in the words of a colleague. He spent the summer of 1929 on another grant-paid trip to Madrid, mastering the art of cell staining with Professor Santiago Ramón y Cajal, the pioneering histologist and neurologist who had won his own Nobel Prize in 1906 for describing the cellular nature of the nervous system. And he began thinking about the gut. In 1931, Florey took his investigations to the University of Sheffield, which had offered him the Joseph Hunter Chair of Pathology. Four years later, he returned to Oxford.

By then, the Dunn School had been open for eight years, led by George Dreyer, a brilliant pathologist who had made his name ana-lyzing diphtheria toxin. Specifically, and following in the steps of

Paul Ehrlich, who had established the standard for measuring the *toxin*, Dreyer standardized the agglutinating power of blood *serum* into a single numerical value.

Standard setting is the sort of unexciting but vital scientific research that tends to be scanted in most histories. It was, however, hugely important, and about to become more so, as pathology and pharmacology became a matter of testing large numbers of different compounds: the more, the better. Consider the hundreds of azo-plus-sulfa side chains that were, at virtually the same time, being produced by Klarer and Mietzsch at Bayer's tropical medical group. The ability to compare samples using a single number was critical, and Dreyer understood this better than anyone in Britain. Among his notable achievements while at the Dunn School was building Britain's first Standards Laboratory, a repository for pure samples of many pathogens.

Dreyer died in 1934, which opened up what, in the ornate language of Oxford, was known as a "Statutory Professorship"—a position within the university that was, by tradition, administered by one of Oxford's thirty-nine residential colleges.* For the Dunn School, that was Lincoln College, whose rector was importuned by a number of dignitaries supporting Florey's candidacy. They included an old mentor—Charles Sherrington—and a new one: Edward Mellanby, the pharmacologist who had discovered vitamin D and demonstrated its relationship to the deficiency disease known as rickets. Mellanby had been one of Florey's supporters on the Sheffield nominating committee, and had since become even more prominent as secretary of Britain's Medical Research Council, which had been founded in 1913 as a publicly funded agency charged with financing medical research, initially on tuberculosis, but from 1920 on, on all diseases. In 1934, those funds were still extremely hard to come by, particularly as compared to the investments being made by the German chemical conglomerates, but the MRC did have one advantage that they lacked:

---

* For those who care: The number, ever since the 2008 merger of Green and Templeton Colleges, is now thirty-eight.

a Royal Charter that authorized collaboration between researchers in academic settings like the Dunn and those in the chemical and pharmaceutical companies.

It was clearly the right moment for such collaboration, and the Dunn was the right place. It remained to be seen whether its new leader—Florey was named director in December 1935—had the right research agenda. It's unclear when, or why, Florey started a series of investigations into the well-known but poorly understood phenomenon that the wall of the gastrointestinal tract was impermeable to bacteria—why potential pathogens didn't infect the wall itself. A number of theories were currently popular, and one of them was the presence of the antibacterial compound lysozyme, discovered by Fleming eight years before.

Florey had been interested in lysozyme ever since, and had even gotten a grant to extract it in a pure form from egg whites. The need for purifying lysozyme led directly to one of the most important decisions in the history of medical research, a decision not about technique or theory, but personnel. Florey, like Domagk, needed chemists to give him the raw material (or, more precisely, the *refined* material) for his experiments. Even while at Sheffield, he had been begging the Medical Research Council for just such a collaborator, to purify Fleming's lysozyme and identify its substrate: the molecular component on which the compound did its work. At Oxford, though, he had considerably more to offer, and, in 1936, he finally got his chemist: E. A. H. Richards, from one of the world's most innovative organic chemistry departments, Oxford's Dyson Perrins Laboratory.* By 1937, Richards had succeeded where Fleming had failed, producing lysozyme in pure form. Even more consequentially, in the same year Florey charged another staff chemist with the job of identifying lysozyme's substrate. The staffer was a Jewish émigré from Germany named Ernst Chain.

---

* Like the Rhodes Scholarship, and Dunn itself, the Dyson Perrins Laboratory was a twentieth-century monument to the philanthropic vanity of Britain's seemingly inexhaustible supply of nineteenth-century tycoons. In this case, the money for the lab was a bequest from the grandson of William Perrins, the originator of the secret recipe in Lea & Perrins Worcestershire Sauce.

Certainly, Chain's background seems, on first glance, to have been wildly different from Florey's. Born in Berlin in 1906, the son of a German mother and a father who had emigrated from Russia's "Pale of Settlement"—the region in eastern Russia that, from 1791 to 1917, had permitted permanent Jewish residency—to Germany, where he studied chemistry, changed his name, and opened the Chemische Fabrik Johannisthal Adershof, a factory that produced pure elements like copper and nickel for industrial use.

This hides more than a few similarities. Florey's father, too, was an immigrant, though from England to Australia. Both fathers took the entrepreneurial path followed by immigrants everywhere, in John Florey's case starting a business manufacturing boots. Both fathers died while their sons were still at school, Florey's in 1918, Chain's a year later, a financial hardship for both families. The death of Chain's father left his family—including Ernst's mother and sister, Hedwig—somewhat strapped, though not so distressed that Ernst could not take advantage of Germany's extraordinary educational system; and in 1927, he graduated from Friedrich-Wilhelm University (now Humboldt University). In 1930, he added a D.Phil. from the Institute of Pathology at Berlin's Charité Hospital, the alma mater of more than half of Germany's Nobel Prize winners, including Emil von Behring, Robert Koch, and Paul Ehrlich. In April, with his new degree and, as he later recalled, £10 in his pocket, he left for England, leaving his mother and sister behind.

Within two years, Chain was publishing papers in scholarly journals and, despite a passport endorsement that should have forbidden him from accepting "paid or unpaid employment," had joined the chemical pathology lab at University College Hospital. In May 1933, despite that pesky business about employment—he was then living on a stipend of £250 annually from the London Jewish Refugees Committee and Liberal Jewish Synagogue—he got a job in the Department of Biochemistry at Cambridge, working for his own mentor: Frederick Gowland Hopkins, who had been awarded half of the 1929 Nobel Prize in Physiology or Medicine (for the discovery of vitamins) and had been, since 1930, president of the Royal Society.*

And, like Florey, Chain was ambitious, confident, and resourceful. In the words of one biographer, he provided an "inexhaustible flow of ideas and suggestions for overcoming difficulties," which would have made him a desirable colleague but for an unhealthy dose of arrogance and a habit of condescending to others, aspects of his personality he was either unable or unwilling to hide. Unsurprisingly, this was, again as with Florey, the cause of frequent irritation among his coworkers. They put up with him, and even sought him out, for his excellent experimental technique, and for a truly remarkable, near-photographic memory. Chain's ability to summon every pertinent reference in the scholarly literature of biology and chemistry without recourse to a library made him the 1930s equivalent of a laboratory connection to the Internet. A dozen different colleagues would later recall his ability not merely to quote page and volume from relevant journal articles, but to quote text verbatim, even citing where on the page the important passage would be found.

Chain had another advantage in the lab. Long before he had become a scientist, he was a piano prodigy, good enough to give concerts in Berlin in his teens, and one who maintained his technique with constant practice. He was gifted enough that at least one account has him visiting Buenos Aires on a concert tour in 1930; certainly, as late as 1933 he was so torn between biochemistry and music that he interviewed for a job in the BBC's orchestra.

Chain's musicianship gets mentioned frequently in biographies, often as evidence of the "artistic temperament" that made him a creative experimentalist. Far more valuable, though, for the laboratories of the

* Since 1885, presidents of the Royal Society have served for five-year terms. In between Florey's mentor, Charles Sherrington (1920–1925), and Hopkins (1930–1935), the president was another Nobel laureate: the physicist Ernest Rutherford. In fact, from 1915 to 1990, fifteen consecutive presidents of the Royal Society were Nobel Prize winners . . . and the streak was broken only with the election of Sir Michael Atiyah, a mathematician whose discipline is unrecognized by Alfred Nobel's trust.

day, was Chain's combination of a pianist's muscle control and hand-eye coordination. Steady hands are as valuable to a chemist as they are to a stage magician; the simple experimental technique of titration depends on adding one liquid to another, a droplet at a time, in order to observe the beginning and end of a chemical reaction in a precisely calibrated vessel. Growing crystals by diffusing one solvent into another has to be done slowly and steadily enough to keep the boundaries between them distinct. Small wonder that Chain's colleagues were as likely to recall his deftness with beaker and pipette as they were his memory.

Credit: Wellcome Library, London

The Dunn School team: Howard Florey (*back row, second from left*) and Ernst Chain (*back row, second from right*)

Florey didn't have Chain in mind when he made recruiting biochemists a priority for the Dunn after he took over in 1935. His first choice for the job was another of Frederick Gowland Hopkins's Cambridge protégés, Norman Pirie, a Scottish biochemist and virologist. Pirie was either uninterested or unavailable, and it was Hopkins who suggested Chain. "I find his biochemical knowledge is more than merely adequate . . . he has really become a well-qualified biochemist. . . ."* Chain was interested in the job. As he would later write, his "principal motivating principle . . . was always to look for an interesting biological phenomenon which could be explained on a chemical or biochemical basis, and attempting to isolate the active substances responsible for the phenomenon and/or studying their mode of action."

Of course, he wrote that long after he had departed Oxford as one of the most famous scientists in the world. When he arrived there, his priority was less on the disinterested search for scientific explanation and more on, well, the toys. After training in the state-of-the-art facilities at Berlin's Charité Hospital, he had been regularly, and loudly, disappointed at the quality of the equipment available at Hopkins's lab in Cambridge. So, when he arrived at the Dunn, and Florey's chief technician, Jim Kent, escorted him to his lab and Chain saw—through the window of the neighboring Dyson Perrins lab—a Soxhlet extractor (a fairly rare piece of lab apparatus designed to extract lipids from solids, and used for all sorts of purifying exercises), his eyes grew as large as a child's visiting a chocolate factory. Asked by Chain if the Dunn School used them, Kent said he believed they had one, to which Chain said, "One! I shall want ten!"

If only. The Dunn was well equipped only by the standards of other British laboratories, and Florey's group still depended on the unpredictable largesse of individuals and foundations, and modest subsidies from the Medical Research Council. In the middle of a worldwide

* Of Chain's downside—his Jewishness—Hopkins continued: "I feel that if his race and foreign origin will not be unwelcome in your department, you will import an acceptable and very able colleague in taking him. Incidentally I have found that his remarkable genius as a musician has made him acceptable in certain social circles. . . ."

depression, none of them was what one might call generous. When Chain arranged for enlarging and modernizing the lab's refrigerator— it had been manually operated by a steward, who turned on the compressor when he thought it was getting too warm—and exceeded the budget by £15, it "caused a terrific upheaval, and Florey never forgot this incident and reminded me of it until I left the Institute."

Over the course of the next year, Chain worked diligently (and parsimoniously) on everything from snake antivenin to protein proteolysis, despite some disabling but undiagnosed diseases that caused him bouts of such severe depression and anxiety that he began keeping a health diary, where, in August 1935, he carefully noted his "periodic fear attacks." Part of his anxiety was, no doubt, financial; with the title of departmental demonstrator, he was paid only £200 annually, and frequently enough needed to get advances on even this modest sum, which he supplemented with the occasional grant.

By the beginning of 1936, Chain had started another research program, this time on skin cancer, and needed an apparatus for measuring the very small amounts of oxygen uptake during the metabolism of cancerous tissues. While devices, known as respirators, existed for less granular investigations, Chain needed something on an order of magnitude more sensitive. To design his "microrespirometer," he suggested one of his former Cambridge colleagues to Florey. Florey agreed, but—surprise of surprises—while he could afford the man, he had no money for the needed equipment.

Once again, the Rockefeller Foundation was tapped. Florey wrote to Daniel P. O'Brien, a former circuit rider, now associate director of the foundation's Division of Medical Sciences, begging for money to expand "the chemical aspect of Pathology." Soon after, O'Brien's boss, Wilber Tisdale, visited Oxford, agreed with the request, and approved a grant that allowed Florey to buy "balances, micro balances, vacuum distillation apparatus, etc. to a value of £250." Far more important than the equipment, though, Florey also agreed to hire Chain's onetime colleague, a mechanical genius named Norman Heatley.

Heatley was then barely twenty-five years old, a newly minted PhD from Cambridge, and, luckily for his colleagues, possessed of a very different temperament than both Florey and Chain. Heatley even looked different. Chain was mustached, short, and intense, his head always set slightly forward, as if ready to attack; Florey was square jawed and robust, laconic and brusque. Heatley was tall, elegant, and slender. Where Florey and Chain were confident, verging on—generally passing over into—arrogant, Heatley was diffident, perhaps to a fault: He was known to have said, "I was a third-rate scientist whose only merit was to be in the right place at the right time." He was also unfailingly courteous, so much so that he was regularly shocked by the egotism and ambition regularly on display at the Dunn. Certainly he got on better with Florey than with his putative supervisor, Chain, especially after the latter insisted on a credit in the journal article that described the microrespirometer, despite the fact that, as Heatley recalled four decades later, the device "was wholly my conception and design" and that Chain had demanded credit out of ambitious careerism.

The disputed article, which would eventually appear under the title "A New Type of Microrespirometer" in the *Journal of Biochemistry* in January 1939, isn't a terribly important scientific paper, but it is a gold mine of foreshadowing. First, it revealed Florey's special talent for preserving the morale, and so the productivity, of all his research assets; when Chain claimed that he deserved credit as an author for the paper, Florey was able to give it to him without simultaneously angering Heatley (at least, not very much). In the final article, as directed by Florey, Heatley is the first-named author, but Chain is the last—in the world of scholarly publishing, the second-best spot. (In between Heatley and Chain was a research fellow at the Dunn named Isaac Berenblum, who would become a world-famous oncologist after emigrating to the new state of Israel in 1949.)

More foreshadowing: On display are not only Florey's careful management style and Chain's ambition, but also Heatley's great talent for building lab equipment out of spare parts and discards. Describing the magnetized iron balls needed to mix the droplets under investigation, he wrote, "Steel-bearing balls, 1/16 in. in diameter, are given several coats of Bakelite varnish . . . the balls are then heated to 100° in paraffin wax for some minutes, the surplus wax being removed by rolling the balls on hot filter paper. . . . They are then rolled in the palm of a warm, but clean and dry hand with some well washed kaolin. . . ."

The article was the first time the peculiar mix of talents of the Dunn team stood revealed. It also marked, or rather caused, a permanent breach in the relationship between Heatley and Chain. From that moment forward, at Heatley's insistence, and with Florey's tacit approval, all communication and direction for the young man from Kent would come from the Dunn School's director, rather than its chief biochemist.

By this time, Chain had plenty of other subjects to keep him busy, most especially the one that had been Florey's reason for bringing him—and E. A. H. Richards—to the Dunn in the first place: finding the substrate for lysozyme. In 1937, assisted by one of that year's Rhodes Scholars, a medical student from Missouri named Leslie Epstein, Chain had found his answer: The substrate was determined to be a polysaccharide, which meant that lysozyme was a polysaccharidase, whose (mild) antibacterial action was that it broke down the polysaccharides (some of them, anyway—*E. coli* alone makes more than two hundred different polysaccharides) that coat the cell walls of bacteria. Epstein had found the subject for his thesis: "The Actions of Certain Bacteriolytic Principles."

Around the same time, Florey found Alexander Fleming's 1929 paper on penicillin.

No one knows precisely how the paper came to the attention of the Dunn investigators, or even which of them first read it. To his death, Chain was adamant that Florey never thought about penicillin until Chain suggested it. "Something seemed to click in my [i.e., Chain's] mind" after reading Fleming's paper. Florey was equally insistent that he had brought it up to Chain. All that can be known for certain is that, during the preceding eight years, virtually no other researcher had cited Fleming's work.

Fleming's discovery was a classic dead end: an interesting compound that was so unstable that even its discoverer couldn't reliably produce it for future experiment, nor could anyone else. Though

Florey's predecessor at the Dunn, George Dreyer, had been intrigued enough by Fleming's mold to secure some for the pathology lab, he had done so for dozens of potentially interesting compounds, and no one had been any more fortunate than Fleming himself in understanding its actions. No one, that is, until 1937, when Florey and Chain began planning an ambitious survey of *all* the antibacterial substances produced by microorganisms. The planned survey would include dozens of different strains of bacteria, but also fungi, particularly the *Penicillium* molds.

The first experiments of the new survey, though, were still focused on lysozyme and other potential antibacterial substances, since penicillin, in its decidedly impure, "natural" form, was such an unreliable antibacterial agent. In any case, Chain believed penicillin to be a kind of "mould lysozyme," an enzyme that acted, like Fleming's egg-white lysozyme, on bacterial cell walls, but also on pathogens like staph and strep, so research on lysozyme was likely to be applicable to penicillin anyway.

In one well-remembered discussion over afternoon tea, Florey reminded his listeners that penicillin was also notoriously difficult to work with; not only had Fleming been unable to stabilize his own compound, Harold Raistrick, a skilled and experienced biochemist, had no better luck. Chain reacted by saying Raistrick couldn't be a very good chemist, since it "must" be possible to produce it in a stable form. Whether deliberately or not, Florey had challenged Chain.

Chain, who was as competitive as a pit bull, responded, though it is worth a reminder that both he and Florey saw the research as an interesting scientific challenge far more than as a way to add to medicine's therapeutic arsenal. Prontosil and the other sulfanilamides were rightly regarded as revolutionary therapies, and there seemed little need or desire to supplant them.

Chain was already investigating the substance produced by *Pseudomonas pyocyanea* (the bacterium responsible for, among other things, septic shock and a number of skin infections; since the 1880s, extracts of *P. pyocyanea* had been shown to destroy other bacteria . . . and to be highly toxic to mammals), and the somewhat more promising

Bacillus subtilis, a hardy microbe with some demonstrated ability to stimulate the immune system. To them, he added the remaining frozen samples of *Penicillium notatum* that Dreyer had left behind, but his first results were unimpressive. He could study the mold if he had enough, but it was "impracticable to grow the [*Penicillium*] mould and carry out chemical studies simultaneously."

Heatley stepped into the breach. Despite his self-effacing modesty, he was, in the words of Gwyn Macfarlane, a hematologist at the Radcliffe Infirmary and later one of Florey's biographers, "a most versatile, ingenious, and skilled laboratory engineer on any scale, large or minute. To his training in biology and biochemistry, he could add the technical skills of optics, glass and metalworking, plumbing, carpentry, and as much electrical work as was needed." Most important of all: "He could improvise—making use of the most unlikely bits of laboratory or household equipment to do the job with the least possible waste of time."

When he was drafted to increase the yield of the antibacterial substance produced by the *Penicillium* mold, he knew relatively little about it, except that the fungus grew adequately on agar, but did best in shallow vessels, no more than 1.5 centimeters deep. At that depth, the branchlike mycelia of the mold could grow above the surface of the agar, and then dry out. Once dry, yellow drops of Fleming's "juice" formed on the dried-out mycelia and could be collected using a glass pipette. Even more valuably, other penicillin droplets settled into the agar itself and turned it yellow. The most productive time for penicillin "farming," therefore, was just after the broth turned rich enough to be harvested, but before it became so saturated that the agar couldn't grow another batch. By careful observation, Heatley learned to identify the agar's phase of maximum productivity.

Agar offered a good base for growing *Penicillium,* but continued to produce frustratingly small amounts of Fleming's broth. The stakes were high; only with significant quantities of broth could any investigation proceed, and Heatley knew it. He fertilized it with everything he could find on the Dunn's shelves: nitrates, salts, sugars, glycerol, meat extracts. He dosed the media with oxygen and CO_2. In December 1939, he tried adding brewer's yeast, which improved the yield only slightly, but did cut the time it took for the mold to produce the broth from three weeks to ten days.

It took months before Heatley determined the best recipe. First, he incubated the fungus on a nutrient solution known as Czapek-Dox medium: a stew of inorganic salts, sugar, and agar. Once the mold bloomed, brewer's yeast was added. In days, a film formed on top of the medium, and soon thereafter, green spores of the *Penicillium* would appear. Over the course of ten days, the fungus would grow, after which Heatley drew off the penicillin-laced broth and replaced the growth medium, twelve times if he was lucky, two or three times if not.

This provided sufficient amounts of broth, but did little to gauge its strength. Though Fleming and those who followed had been able to demonstrate the antibacterial activity of the liquid extracted from mold, what they had wasn't penicillin, but a broth of which penicillin was a component. How much of the liquid was biologically active? No one knew. Heatley needed a yardstick for measuring antibacterial activity, and once again he found an ingenious solution: He cut disks out of the bottoms of Petri dishes and replaced them with glass tubes, creating concave indentations in the center of the vessels, into which colonies of bacteria were introduced. He then added measured amounts of the cultivated yellow broth, and noted the size of the bacteria-free halo around each cylinder. The bigger the halo, the more potent the compound.

If Heatley was legendarily resourceful in making equipment for pennies, Florey was no less so in collecting the pennies themselves, wherever they could be found. The use of the word "pennies" is not accidental; to call the Dunn experimental program impoverished is to flatter. At the 1938 Physiological Congress in Zurich, Florey buttonholed Edward Mellanby to beg the Medical Research Council for what seems, in retrospect, a ridiculously tiny sum: £600. Even when he got it, it was scarcely enough. At one point, Florey told Chain that the lab had completely exhausted its funds, and that he must stop ordering everything, up to and including glassware.

Though the Medical Research Council finally agreed to renew a portion of the lab's grant for 1939—for Chain, £300 a year for each of the next four years, plus an additional grant for expenses of £250 annually through 1940—it kept it just north of starvation. When Florey learned that Oxford was proposing to cut the operating budget they supplied to the Dunn because a new university heating plant would lower the laboratory's utility costs, Florey wrote back, "I have struggled to keep the place warm on money I ought to have devoted to research."

Matters nearly came to a head in the summer of 1939, when the grant from the Medical Research Council that paid for Chain's research on the penicillin project was about to expire. Florey might not have loved Chain's company, but he recognized his value and was determined to find the money needed to keep his biochemist fed, housed, and not too surly for work. Their grant application not only identified fungi as a promising source for antibacterial compounds, but announced the status of their experiments on Alexander Fleming's nearly forgotten eleven-year-old discovery.* For the first time, penicillin was an explicit part of their research program.

Florey was taking no chances. He sent a virtually identical grant proposal off to the newly established Nuffield Provincial Hospitals Trust (another charity built on a wealthy man's will, this time a bequest from William Morris, who had made his fortune selling Morris Garages sports cars—MGs), and another to the usual suspects at the Rockefeller Foundation, though by a circuitous route. The original request was sent to the Rockefeller offices in France on November 20, and was then forwarded to the New York office ten days later, where

---

* Perhaps to cover all bets—researchers, then and now, live from grant to grant, and are obsessively careful to include any information that might interest a patron—the application also promised to expand on an exciting new discovery from the French-born American microbiologist René Dubos, then working at the Rockefeller Institute for Medical Research: tyrothricin, an antibacterial compound produced by a member of the class of soil-dwelling bacteria then known as the actinomycetes. For more about Dubos and the actinomycetes, see Chapter Six.

the director of scientific research, Warren Weaver, wrote back, "The application of Florey appeals to me, but I seriously question whether a three-year grant is justified under present circumstances. . . ."

The "present circumstances," were, of course, the Second World War, which had commenced with Germany's invasion of Poland on September 1, and within days, declarations of war from France and Britain. By the time Weaver received the Dunn School's grant application, Poland had been divided between Germany and the Soviet Union, the United States had passed the Neutrality Acts (which allowed France and the United Kingdom to buy arms), and the estuaries of the Thames had been mined by U-boats. The threat of war had already had an impact at the Dunn; Chain, as a refugee perhaps more fearful than most of a German invasion, and more grateful to Britain for offering sanctuary, had volunteered for a Red Cross Certificate in First Aid, and, after becoming a British subject in April 1939, joined the Oxford City Council Air Raid Precautions Department. Heatley was unable to leave England for a fellowship in Copenhagen, and—at Florey's behest—stayed in Oxford.

Initially, the transition to open hostilities would shave budgets for research, particularly of the "interesting scientific challenge" sort. After some back-and-forth, including reassurances from Weaver that the grant would be renewed as long as the Dunn team showed progress, the money started flowing again, though, as always, through a very narrow straw. Though Florey's original request to the Medical Research Council for studying penicillin as a therapeutic agent in vivo was for a mere £100, the MRC actually came through with only £25. Luckily, on February 19, 1940, the Dunn team learned that the Rockefeller grant had been approved, with the first payment scheduled to arrive March 1.

Tiny budgets notwithstanding, the Dunn research was progressing, and progressing rapidly. By March 1940, Heatley's methods had improved yields so much that instead of providing Chain with a milligram of broth at a time, he could make a hundred times as much.

However, the trick of extracting penicillin from the broth in a

stable form, which had eluded Raistrick and others, was still undis-
covered. Fleming had tried to separate the active ingredient in his
mold juice using a simple chemical technique: Dissolve the stuff in
ether, which would evaporate quickly, leaving behind concentrated
penicillin. It had failed almost completely. Harold Raistrick, a far
more skilled chemist, used a method of separation known as liquid-
liquid extraction: Pour the mixture into a "separatory funnel," a piece
of apparatus that looks like an inverted teardrop, with a funnel at the
top, and a stopcock at the bottom, with a flask underneath. In the top
chamber, liquids separate into different layers based on densities,
with the heavier "aqueous" layer, containing ions, or charged particles, at the bottom, and a neutral, uncharged "organic" layer at the
top. Shake vigorously, open the stopcock, let the bottom layer flow
out, and you have your extract.

The key, then, was to add a charge to the molecules that composed
the desired portion of the broth.* There are a couple of ways of charging a neutral substance, but one of the most effective is to acidify it,
since what makes acids acidic is a freely given positively charged hydrogen atom: a proton. Give a proton to a neutral compound, and it
becomes charged. The greater the charge, the greater the amount of
the previously neutral substance that heads to the bottom of the separatory funnel. This was Raistrick's strategy. By adding acidified
ether to the solution, a little at a time, he was able to separate the
mold juice from superfluous fluid, leaving behind a concentrate of
about one-fifth the volume with which he began.

However, when he evaporated the ether (easy enough; it's a very
volatile substance) expecting to find a concentrated form of penicillin
in the residue, it had completely vanished: None of the antibacterial
activity that Fleming had identified in the mold juice remained.

In the intervening years, no one had been able to improve on

---

* This was a *very* small proportion; though neither Raistrick nor Fleming could
know this, the concentration of penicillin in the original mold juice was less than one
part per million.

Raistrick's method. This meant that, despite Heatley's resourceful-ness in cultivating *Penicillium*'s precious broth, no one knew how to convert his harvest of broth into a stable source of penicillin.

Enter Heatley, again. He first tried to stabilize the compound at dif-ferent temperatures and pH levels, slowly adding alkali, or base, salts to the mixture to get it back to neutral, but it wasn't a very practical method, the equivalent of baking a soufflé over a campfire. His second idea, what he later called "laughably simple," reextracted the com-pound from the acidic ether solution into a neutral medium—water—by taking another trip through the separatory funnel, and then gave it another charge, this time by exposure to a base. On March 19, 1940, he did just that, filtering the mold broth through parachute silk (in order to remove solid particles), then mixing it with ether, which caused it to layer and separate. The ether-plus-penicillin mixture was then mixed with alkaline salts, dissolved in water, and separated again; only this time, the penicillin was in the aqueous layer. And, unlike the ether-plus-penicillin mixture that had lost all its potency for Raistrick, the new mix was stable, even after eleven days at room temperature. A source for experimental quantities of far purer penicillin—though the term is used loosely; while the concentration of penicillin in Fleming's broth was only one part per million, Heatley's first batch was still no better than .02 percent pure—had been found.

And Chain was ready to experiment with it. Almost as soon as he received the purer extract of penicillin, he directed John M. Barnes, one of the few researchers at the Dunn licensed by the Home Office to perform animal tests, to inject the entire stock of penicillin extract, 80 milligrams, into the abdomens of two mice.

It isn't clear what he expected in the way of a reaction. Over the pre-ceding year, Chain's supposition about the nature of penicillin—that it was a complex protein molecule, probably an enzyme, like lysozyme—had been fading. Though the explicit aim of the 1939 grant was the "preparation from certain bacteria and fungi of powerful bactericidal enzymes, effective against staphylococci, pneumococci, and strepto-cocci," by the beginning of 1940, Chain had completed a number of experiments that proved penicillin couldn't be a "bactericidal enzyme," or indeed a protein of any sort. In one experiment, penicillin dialyzed— broke into component parts—when forced through cellophane tubes— which is something that proteins, because of their size, don't do; in Chain's words, his "beautiful working hypothesis dissolved into thin air." If penicillin had been a protein, the mice would have exhibited an immune response: swelling, perhaps, or inflammation. But they didn't. The mice might as well have been given saline solution; there was no impact. Penicillin definitely wasn't any sort of protein.

The bad news was that Chain hadn't found a complex molecule, but a relatively simple one, though, again in Chain's words, "it became very interesting to find out which structural features were responsible for the instability. It was clear that we were dealing with a chemically very unusual substance." The good news was that the mice tolerated it almost completely, which suggested that, unlike almost every other promising antibacterial compound, it seemed safe. Even better: the extract, once excreted in urine, was a) nearly as brown as the compound injected, and b) still strongly bactericidal.

The Dunn team was on the verge of isolating a substance that killed pathogens without damaging their hosts. Which made it *highly* interesting medically.

Howard Florey made penicillin the Dunn School's number one priority. Two months after Chain's first test, at 11:00 A.M. on May 25, 1940, Florey infected eight mice with *Streptococcus pyogenes*, the pathogen responsible for infections like strep throat, impetigo, erysipelas, and even the flesh-eating nightmare, necrotizing fasciitis. At noon, two mice were given 10 milligrams of penicillin, and another two were given 5 milligrams. Follow-up doses were given at 4:15, 6:20, and 10:00. Just before midnight, as Heatley recorded in his lab notebook, "one mouse got up and staggered about for a few seconds, then fell down, twitched once or twice, and was dead." By 4:00 A.M. on May 26, all four of the controls had died.

All four of the treated mice survived.

The following day, as the Battle of France raged, nearly seven hundred British ships and small craft started evacuating what Prime

Minister Winston Churchill called the "root and core and brain of the British Army" from Dunkirk. It would be four years before they would return, on June 6, 1944: D-day. When they began the liberation of France, the most important and valuable medicine used by their medics and field hospitals would be the same substance that had saved four of the Dunn School's mice.

FOUR

"The People's Department"

T he Dunkirk rescue was the best news that Britain would experience for months, as the seemingly unstoppable Wehrmacht conquered
Belgium, Norway, the Channel Islands, and—on June 25—France. The
Dunn team was acutely aware of the German threat, but even more focused on the progress of their own investigations. The success of the May
25 experiment launched the already highly energetic Howard Florey, and
everyone else at Dunn, into another gear. Within days, more successful
experiments with mice followed, revealing the potency of even highly
diluted concentrations of penicillin.

It also exposed a pressing need for more staff. Though the mice
experiments showed the effectiveness of the penicillin broth, they
revealed little about why it worked. Bacteriologists Duncan Gardner
and Jean Orr-Ewing were brought on board to investigate the mechanism by which the compound was performing its magic. Florey himself, along with Jim Kent, planned and executed a series of experiments
comparing the results of different dosages of both *streptococci* and
penicillin at different stages of infection.

The chemical investigations were accelerating. The Dunn team
needed more penicillin at more powerful concentrations, and that
could only happen with a better understanding—any understanding,
really—of penicillin's chemical structure. A twenty-seven-year-old
chemist named Edward Penley Abraham was assigned to work with
Chain on the daunting problem of purification.

Abraham and Chain were a well-matched team. They used a variety of purification techniques aimed at improving the compound's

potency, as measured by the ratio of the active ingredient to the rest of the solution: then as little as 0.5 milligrams per liter, less than one part in a million. The most effective technique turned out to be freeze-drying: First, dissolving the penicillin with dry methanol, diluting it with water, then freezing the liquid. Second, exposing it to a high vacuum to sublime the solvent, transforming it directly from a solid to a gas. Third, after all the sublimable solvent has been removed, separating the rest of it by exposure to high heat. The real eye-opener for the two chemists was the discovery that, even when diluted to one part in a million, penicillin still stopped bacteria from growing, which made it at least twenty times more powerful than the strongest sulfa drugs.

This was promising and frustrating in equal measure. Penicillin might be the most powerful and effective medicine ever discovered, but it was also one of the most difficult to produce. The lack of raw material was the most daunting bottleneck for further research. How, then, to produce more? Despite the grants from the Rockefeller Foundation and the Medical Research Council, Florey was still scrambling for money, and now needed larger-scale production resources.

One intriguing possibility would be to enlist Britain's pharmaceutical industry, whose companies, though generally far smaller than Germany's huge conglomerates, nonetheless possessed manufacturing expertise, investment capital, and, especially, a powerful interest in any compound with the potential to treat infectious disease. Glaxo, founded in New Zealand in 1880, had expanded into the United Kingdom less as a pharmaceutical company than as a manufacturer and processor of milk fortified with vitamin D;* by the 1930s, they were the country's largest producer of nutrition products. Beecham's Pills were Britain's most popular laxative, and the foundation of Beecham Limited. Imperial Chemical

---

* Vitamin D occurs naturally in oily fish and some animal organ meats, such as liver. Trace amounts are sometimes found in full-fat milk, but not enough for dietary sufficiency. Since milk was so widely consumed by the children at greatest risk for rickets, the deficiency disease associated with a lack of vitamin D, fortifying milk by a number of different processes had become a common practice since the early 1930s.

Industries, the result of an I. G. Farben–style merger between competing companies (including Nobel Explosives and the British Dyestuffs Corporation), had been, since 1926, by far Britain's largest chemical company, but, unlike the Germans, had little or no interest in pharmaceuticals. The small chemical manufacturing firm Kemball, Bishop and Company showed some interest, visiting the Dunn in March 1940 accompanied by Sir Henry Dale, president of the Royal Society, but its resources weren't up to the task.

More promising was Burroughs Wellcome, then Britain's most technologically sophisticated pharmaceutical company. The company had been founded by two American expatriates, Henry S. Wellcome and Silas M. Burroughs, graduates of the Philadelphia College of Pharmacy, who had decamped to London and opened for business in 1880. Sixteen years later, the forward-thinking Americans had built Britain's first industrial biological research facility, the Physiological Research Laboratories, largely to produce the company's own version of Robert Koch's antidiphtheria serum.* When two chemists from Burroughs Wellcome visited the Dunn in July 1940, they politely declined Florey's suggestion that they take on the challenge of purifying and producing penicillin.

With no other takers, the Dunn would have to produce the penicillin it needed for research in-house, making up in ingenuity what was lacking in money. What this meant, in practice, was throwing the job to Norman Heatley.

Heatley's first challenge in producing sufficient quantities of the precious substance was a shortage of properly sized vessels for growing

---

* The early company had prospered by adapting machinery initially invented for the manufacture of ground and compressed graphite for pencils for use in producing consistently pure pills, typically alkaloids like morphine and codeine. This is why it was a Burroughs Wellcome marketing genius who first coined (and trademarked) the word "tabloid" as a combination of tablet and alkaloid. In the fullness of time, the term was transmuted into the common name for a newspaper with a policy of compressing lengthy news stories into something briefer, but to this day the company's Web site boasts of Burroughs Wellcome's ownership of the term.

the mold itself. Faced with extraordinary difficulty in procuring appropriate glassware (or anything; Florey was so penurious that he had the Dunn's elevator shut down in order to save £25 annually), Heatley turned to larceny. Pie trays and baking dishes mysteriously vanished from the Dunn's kitchens. Sixteen bedpans likewise disappeared from the Radcliffe Infirmary, to reappear in the pathology department's labs.

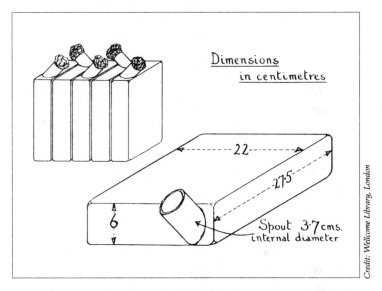

The modified ceramic bedpans, used by the Dunn School team to grow penicillin broth

Over the course of 1940, Heatley continued to improve his technique for extracting penicillin from the broth produced in his bedpan factories. The most effective process he found depended on a truly remarkable machine, able to mix ether with penicillin, acidify it, and separate the concentrated broth.

The homemade apparatus in operation resembled nothing so much as a Rube Goldberg cartoon.

1. Three bottles—of broth, ether, and acid—are held upside down in a frame, until

2. The glass ball stopper in the bottle containing broth is moved aside; liquid flows

3. Into a glass coil surrounded by ice. Once cooled, the acidified liquid combines with acid from bottle number three and is jet-sprayed in droplets

4. That arrive in one of six parallel separation tubes. Meanwhile

5. The stopper on bottle number two, containing ether, is moved aside, releasing ether into the bottom of the whole arrangement. The filtrate in the separation tube is sprayed into a tube of ether rising in a four-foot-long tube. As penicillin has a chemical affinity for the ether, it transfers into that tube, leaving the remaining components of the original broth behind, to be drained out.

6. Then, the penicillin-plus-ether (later acetate) solution is introduced into another tube, with slightly alkaline water. The penicillin-plus-water mixture—about 20 percent of the volume of the filtered broth that started the whole rigamarole—was drawn off.

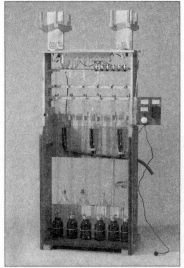

Credit: Science & Society Picture Library

The filtration machine built by Norman Heatley to purify penicillin

Remarkably, Heatley's collection of discarded junk and baling wire—the hole in the glass needed to produce the right-sized droplets was made by Heatley pushing the point of a sewing needle through hot glass; a cast-off doorbell rang when each bottle was filled—could turn about twelve liters of broth into two liters of decidedly impure penicillin in an hour. A quantity of penicillin sufficient for experimentation (on mice, at any rate) had been guaranteed.

Florey decided it was time to go public. The *Lancet* of August 24, 1940, contained half a dozen articles. Two of them—topically enough, given the Blitz—were on treating blast injuries to the lung. Another article was on meningitis; one described the orthopedic trauma known as "locking wrist." Right in the middle, though, was the world changer: "Penicillin as a Chemotherapeutic Agent," by E. Chain, H. W. Florey, A. D. Gardner, N. G. Heatley, M. A. Jennings, J. Orr-Ewing, and A. G. Sanders. (Florey, having already suffered through the sniping between Chain and Heatley over authorial credit, deferred to the alphabet.) The first line of the article reads, "In recent years interest in chemotherapeutic effects has been almost exclusively focused on the sulphonamides and their derivatives. There are, however, other possibilities. . . ."

The possibilities were detailed in the following two pages, including results from the five key studies performed at the Dunn since March, in which as many as seventy-five mice had been exposed to pathogens as varied as *staphylococcus, streptococcus*, and *clostridium*: "During the last year methods have been devised here for obtaining a considerable yield of penicillin, and for rapid assay of its inhibitory power. From the culture medium a brown powder has been obtained which is freely soluble in water. It and its solution are stable for a considerable time and though it is not a pure substance, its antibacterial activity is very great. . . ."

Eleven years and five months after Alexander Fleming had published "On the Antibacterial Actions of Cultures of a Penicillium," his discovery had finally been revealed as more than just a Petri dish curiosity: "The results are clear cut, and show that penicillin is active *in vivo* against at least three of the organisms inhibited *in vitro*."

Even before the *Lancet* article appeared in August, work at the Dunn had been proceeding on three intersecting tracks. Norman Heatley and the technical staff continued to improve the processes by which penicillin-rich broth was grown, and from which the active ingredient could be extracted. The biochemical team, primarily Chain and E. P. Abraham, were subjecting the compound to a series of experiments intended to establish its structure. And the bacteriologists and pharmacologists were designing new ways to test the antibacterial properties of the compound on laboratory animals.

It is a testimony to Florey's great gifts as a scientific administrator that the projects—production, analysis, and effectiveness—all succeeded brilliantly, in spite of the enormous amount of friction produced by the Dunn's collection of strong personalities. The achievement is notable even though it might be said that at least some of Florey's challenges in managing the lab were self-inflicted. By 1940, he had begun an affair with Margaret Jennings, a physician and histologist who had joined the Dunn in 1936 and became indispensable to the Australian both as a lab assistant and as the editor of the entire lab's scientific publications. Which meant that Ethel Florey, who had been assigned to supervise the penicillin team's clinical trials, was having her papers and reports checked for clarity by her husband's mistress.*

By comparison, soothing the sensitivities of Ernst Chain was a challenge scarcely worth mentioning. He and E. P. Abraham knew there was no chance of deciphering the structure of penicillin until it could be crystallized—precipitated into something stable enough to be analyzed.

---

* Margaret Jennings, who would become the second Mrs. Florey after Ethel's death, was, in many respects, the polar opposite of her predecessor, and many of Howard Florey's biographers have made much of it. Margaret was robust where Ethel was both deaf and sickly. She was English gentility (the daughter of a baronet), physically affectionate, and compliant; Ethel was none of the above. But it's also worth remembering that both of Howard Florey's wives were highly skilled scientists, each of whom played a significant role in the most important medical discovery of the twentieth century, which cannot be a coincidence.

Abraham and Chain were still struggling with the structural chemistry when the *Lancet* article was published. Aware of this, the journal's editors appended a note reading, "What [penicillin's] chemical nature is, and whether it can be prepared on a commercial scale, are problems to which the Oxford pathologists are doubtless addressing themselves," which reads as a bit of an understated scold, as if the Dunn team were keeping at least some of their work secret for now, possibly in order to ensure that they would receive full credit for the discovery.

They were on to something. On September 2, Alexander Fleming paid the Dunn a surprise visit—more a surprise for some of the team than others; Chain evidently thought Fleming was dead—to find out what had been done "with my old penicillin." The maneuvering for credit was well under way.

Anyone familiar with the adage about success having a thousand fathers might have predicted what would ensue. The *Lancet* article had ended, courteously enough, with a footnote from the authors thanking the Nuffield Trust, the Medical Research Council, and the Rockefeller Foundation for their support. It did not have the intended effect. In a foreshadowing of the looming disputes about credit and recognition that would attach to the penicillin discovery, Edward Mellanby scolded Florey about showing more gratitude to the Rockefeller Foundation than his own MRC: "I shall be surprised if the Rockefeller Foundation are supporting the work to anything like [the] extent [of the MRC support; he was unaware of the 1939 Rockefeller grant] . . . if you have a good thing in your own country, you might as well give it proper credit and not follow those people who, in cases of research, find it more convenient to give foreigners boosts than their own colleagues. . . ."

Sometime shortly after the article's publication, Florey received a report from Ernest Gäumann of the Swiss Federal Institute of Technology informing him that he had been approached by the Basel-based concern known as Chemische Industrie Basel, or CIBA, to help purify and manufacture penicillin for the company. More alarming, at least to Florey, was the other news from Switzerland: German researchers were eager to examine any available samples of penicillin.

It was a real dilemma. On the one hand, penicillin promised to be a spectacularly important scientific and medical advance, and one of the canons of twentieth-century science was that such discoveries should be shared as widely as possible. Moreover, there was no legal ground for refusing knowledge about therapies, even when the therapeutic information about penicillin was still the very definition of provisional. On the other hand, Britain was literally fighting for its life. Sharing information about a drug that might accelerate the recovery of wounded German soldiers seemed a lot like giving aid and comfort to the enemy. Florey seems not to have tortured himself very long about competing loyalties. He immediately wrote to Edward Mellanby, saying it was "very undesirable that the Swiss and hence the Germans should get penicillin, and I think it would be well worth while to issue instructions to the National Type Collections not to issue cultures of *Penicillium notatum* to anyone with [a] possible enemy connection, and to send a letter to [Alexander] Fleming to the same effect."

Mellanby's reply: "I sympathize with your position, but I do not see how the Medical Research Council can ask their National Type Collections to restrict their dispatch of special cultures . . . to a neutral country like Switzerland." He went on: "If the sulphonamide compounds had not proved to be so efficacious, I think you might have had a strong case [but] although I do not doubt that penicillin may prove to be superior to the sulphonamide compounds, I have difficulty in believing that this superiority is so great that national interests dictate the withholding of publication."

By January 1941, Heatley's penicillin factory had produced enough penicillin to move from testing the stuff on mice weighing 20 grams or so to 150-pound humans. The purpose was as much to discover whether the compound was dangerous as to test whether it was efficacious. Biologists and pathologists had, after all, discovered dozens of antibacterial compounds that were therapeutically useless because they attacked healthy cells as aggressively as they did pathogens. Florey and the clinical physician he had recruited to administer treatment to human patients, L. J. Witts of the Radcliffe Infirmary, considered who would be the subjects for such a test. Volunteers from Oxford? Someone already at death's door? The Dunn team decided on the latter. On January 17, Elva Akers, a patient at Radcliffe whose cancer was so far advanced that she was given only a month or so to live, volunteered to receive an injection of a tiny amount of penicillin: 100 milligrams.

Given the dozens of mice who had received weight-comparable doses of the drug with no ill effect, the injection ought to have been safe for a human. It was not. Mrs. Akers almost immediately experienced a high fever accompanied by seizures. The reason, however, wasn't the penicillin, but the contaminants that the biochemical team had been unable to segregate from the penicillin itself. This was a side effect of Chain and Abraham's success in separating the penicillin filtrate into different layers; by making one of them relatively pure— up to 80 percent pure—they had also enriched the other layers with a high percentage of impurities, and at least some of them were pyrogens: fever-causing compounds.

Pure penicillin, they quickly learned, was benign. The same separation process that Chain and Abraham were using to decipher its structure could also purify it sufficiently, using a more stringent method of chromatography to eliminate the poisons hitchhiking alongside. Mrs. Akers reacted to a second round of injections with neither fever nor trembling.

Safety, though, wasn't the same thing as therapeutic value. For the next test, the Dunn team needed someone suffering not from cancer, but infection.

They didn't need to look far. The previous September, an Oxford policeman named Albert Alexander had been working in his rose garden when he scratched his face on a thorn. The scratch became infected, first just at the site of the injury, but soon the bacteria that had been so abundant in the soil around the Alexander garden— *streptococci* and *staphylococci*, at a bare minimum—began to multiply and deposit their own toxins in the victim's body. By October, his scalp had become obviously infected, and Mr. Alexander was admitted to the Radcliffe Infirmary, where, despite the use of sulfa-nilamides, the infection spread to his lungs. By February, he had ab-scesses growing in his torso, on his arms, and in his left eye, which he would soon thereafter lose. When Norman Heatley saw him in early February, he noted in his diary that the constable "was oozing pus everywhere."

On February 12, 1941, Mr. Alexander was given an intravenous injection of 200 milligrams of penicillin—still, though his physicians wouldn't know this until much later, less than 5 percent pure—and a follow-up intravenous drip of 100 milligrams every three hours. Af-ter a single day, the eight injections had caused a miraculous improve-ment. Alexander's fever had vanished, he was no longer discharging pus, his face was no longer swollen, and he was able to eat.

The problem was keeping the penicillin flowing at a rate sufficient to maintain the therapeutic effect; it took Heatley's machines days to produce the amount of penicillin Alexander needed every hour. The Dunn team had learned during the mice experiments that penicillin was quickly excreted by the kidneys, while still retaining its antibac-terial properties, so the doctors set up a procedure for collecting Alex-ander's urine after each dose, then carrying it via bicycle from the Radcliffe Infirmary to the Dunn laboratory (a mile and a half each way) in order to extract more of the precious stuff.

Alexander wasn't the only one who needed it. Another patient, a fifteen-year-old boy named Arthur Jones who had contracted a life-threatening infection after a hip operation, was getting a similar course of treatment. The Dunn physicians hadn't known how effec-tive penicillin would be, nor did they have a clue how much was re-quired for a therapeutic dose. By the end of February, the supply of penicillin, even the recycled variety, was exhausted. Arthur Jones survived. Albert Alexander, however, died on March 15, 1941.

The stuff worked—when, that is, there was enough of it. However, no British university lab was equipped to produce the kilogram of pure penicillin Florey estimated would be needed for the next round of clinical trials. Neither was any chemical firm in the Commonwealth.

France was occupied, Germany, Japan, and Italy enemies. Only one place was left.

After the fall of the Soviet Union, it became something of a cliché to describe the United States as the "world's lone superpower." In economic terms, it had already earned the title by 1912. The year that Paul Ehrlich discovered Salvarsan, Germany's gross domestic product was a bit more than $227 billion, just ahead of the United Kingdom's $216 billion. That year, the United States economy was larger than both of them combined: $498 billion. By 1940, the GDP of the just-out-of-the-Great-Depression United States was closing in on a trillion dollars a year.

To be sure, American economic dominance wasn't uniform. Though U.S. steelmakers rolled out 43 million metric tons in 1940, nearly a third of the world's total (Germany's 22 million metric tons earned it only a distant second place), American chemical and pharmaceutical firms were minnows next to I. G. Farben's whale. And while researchers in U.S. universities and industries were well on the way to the dominant position they would assume after the Second World War, Germany's scientific reputation, particularly in physics and chemistry, was still dramatically higher. Though the Nobel Prize is a notoriously imperfect yardstick of scientific achievement, it isn't a coincidence that by 1940 Germany had won thirty-three of the science Nobels. Americans had won twelve . . . three of them, in 1934, for the same discovery: pernicious anemia.

In one area, however, the United States was unmatched: Both the sophistication and productivity of America's agricultural sector was like nothing else in the world. Farms, forests, and ranches still made up nearly 20 percent of the largest national economy in the world.

Which is why, when Warren Weaver of the Rockefeller Foundation visited Oxford on April 14, Florey proposed a visit to the United States, explicitly to find "some American mold or yeast raiser who would undertake a large-scale production of this material for a test, say, 10,000 gallons. . . ." Weaver was convinced almost immediately, and authorized $6,000 in expenses for the trip. Florey needed only one authorization: to leave England. In late April, he wrote to Edward Mellanby asking for help in getting the required wartime exit permits for himself and Norman Heatley, Florey's choice for an expedition devoted to cultivating penicillin in large quantities. A few days later he received this reply: "I have come to the conclusion that the only way that this most important matter can be pursued is for you and Heatley to go to the United States of America for three months."

On June 27, 1941, Florey and Heatley left Oxford by car for a trip to a top-secret airfield, where they boarded a Dutch passenger plane and flew to Lisbon, landing seven hours later. There they met representatives of the Rockefeller Foundation, and three days later continued their journey aboard the Pan American Airways's *Dixie Clipper*. The flying boat discharged its passengers at the Marine Air Terminal of La Guardia Field on the afternoon of July 2. Within hours, Florey was sitting at the head of a conference table in the Rockefeller Foundation offices in Manhattan, explaining the significance of the penicillin experiments to Alan Gregg, the head of the foundation's Medical Science Division. To be fair, American awareness of penicillin had preceded Florey and Heatley. The *Lancet* article had prompted a team at Columbia University College of Physicians and Surgeons to request samples of the compound from Chain and to set up a production line in Manhattan similar to Heatley's in Oxford. By October 1940, in fact, the Columbia team had produced enough penicillin—still the crude and impure filtrate—to inject two humans with it, even before the Dunn team had done the same with Mrs. Akers. A soil scientist at Rutgers College in New Jersey, Selman Abraham Waksman, started some investigations into *P. notatum*,* as did researchers at the Mayo Clinic in Rochester, Minnesota.

Nor were America's small but energetic pharmaceutical firms slow to show interest. Decades before, Parke-Davis had agreed to a cooperative relationship with St. Mary's Inoculation Department; after reading the *Lancet* article, Parke-Davis executives asked Almroth Wright whether he or Fleming could help to secure a sample of peni-

---

* More about Waksman—much more—in Chapter Six.

cillin from Oxford. (This may explain Fleming's surprise visit to the Dunn School the preceding September.) Pfizer, a chemical company headquartered in the Williamsburg section of Brooklyn, that made most of its income from producing the preservative and flavoring compound citric acid, also had a small interest in medicines. So did E. R. Squibb, another Brooklyn-based company and a major producer of surgical drugs like ether. Even more interested was the American branch of the German drug company Merck, which had begun cultivation of *P. notatum* as early as January 1940, and whose president, George Merck, was so well known to Alan Gregg that the Rockefeller Foundation executive proposed a meeting with Howard Florey.

Florey was interested, but he had more immediate business. In July of the preceding year, with the Battle of Britain raging, the Floreys had sent their two children, Paquita and Charles, to New Haven, Connecticut; Howard's onetime Rhodes Scholar companion John Fulton, now the Sterling Professor of Physiology at Yale, had agreed to take them in for the duration. On July 3, Florey headed to Connecticut, intending to surprise his children* and to meet with some like-minded scientists. Fulton was able to perform a dozen different introductions, but none were more significant than those he brokered between Florey and Ross Harrison, chair of the Executive Committee of the National Research Council, which had been responsible for applying "scientific methods in strengthening the national defense" since 1916. Harrison, in turn, arranged an introduction to Charles Thom, a mycologist in the Department of Agriculture's Bureau of Plant Industry in Beltsville, Maryland, and the following week, Florey and Heatley headed for Washington to meet him.

Thom already had a long-standing connection to the world of penicillin research. He was the scientist who had originally corrected Fleming's misidentification of his penicillin-producing fungus—not *P. rubrum*, as Fleming originally had it, but *P. notatum*, using a sample

---

* In this, he would be only half successful; the ten-year-old Paquita was already off to summer camp, and only her five-year-old brother was there to meet his father.

sent to him by his old friend Harold Raistrick.* More important, though, than Thom's previous achievements were his current interests and capabilities. Only a few months earlier, a number of his protégés had been relocated from their own labs in Arlington, Virginia—the War Department had taken the land for building what would become the Pentagon—to the Agriculture Department's largest midwestern lab. As a direct result of their discussions with Thom, on July 12, Howard Florey and Norman Heatley boarded a train leaving Washington, DC's Union Station bound for Chicago, where they would make a connection to the *Peoria Rocket*. Their final destination was the Department of Agriculture's Northern Regional Research Laboratory, in Peoria, Illinois.

In July 1862, Abraham Lincoln signed the Morrill Act, establishing "at the Seat of Government of the United States, a Department of Agriculture."† In late 1864, during his last address to Congress Lincoln said, "The Agricultural Department . . . is rapidly commending itself to the great and vital interest it was created to advance. It is precisely the people's Department, in which they feel more directly concerned than in any other."

Eight decades later, the United States was a considerably less agrarian society, but the USDA was, if anything, even more important to its prosperity. In addition to its programs promoting American food production and providing credit to American farmers, it funded more than forty experimental stations, research centers where farmers, ranchers, and agronomists worked to improve existing agricultural products and practices and develop new ones; provided

---

* Thom's position as a government scientist required him to balance the needs and demands of commerce—his work on two industrially important members of the *Penicillium* family, *P. roqueforti* and *P. camemberti*, made him a hero to American cheese producers—and society: His primary job, almost from the moment he joined the Department of Agriculture in 1904, was enforcement of the provisions of the Pure Food and Drugs Act of 1906.

† Lincoln also signed, on March 3, 1863, the charter authorizing the predecessor of the National Research Council.

ongoing education in agronomy and animal husbandry through its cooperative extension programs; and was the primary enforcer of the provisions of the 1906 Pure Food and Drugs Act, including the penalties for misbranding medicine. Of more immediate relevance to Florey and Heatley, though, were the USDA's four regional labs, the very top of the pyramid in agricultural research.

The four labs—in addition to the Northern Regional Research Lab in Peoria, the USDA operated an eastern lab in Wyndmoor, Pennsylvania; a southern lab in New Orleans; and a western lab in Albany, California—were responsible over the years for literally thousands of innovations, patented and otherwise, from instant mashed potatoes to wrinkle-free cotton. It seems safe to say, though, that the regional labs' finest hour can be dated from the arrival of Howard Florey and Norman Heatley on the *Peoria Rocket* on July 14, 1941.

The Northern Lab looms deservedly large in any history of penicillin— in any history of medicine, really. It was where three of the most urgent objectives in transforming the Dunn discoveries from a laboratory process to an industrial one were achieved: first, the discovery and identification of the most productive strains of *Penicillium* mold; second, a protocol for accelerating the growth of penicillin-producing mold; and third, improvement of the fermentation technique by which the exudate appeared. In traditional agricultural terminology, they were looking for better seeds, better soil, and better cultivation and harvesting.

Better seeds first. Even before Florey and Heatley arrived in July, the Northern Lab's chief mycologist, Kenneth Raper, had sent messages to researchers all over the globe (even to the point of enlisting crews in the U.S. Army's Air Transport Command) requesting that they collect samples of *Penicillium* mold and send them to Peoria. By early 1941, he had started testing dozens of different strains. The most significant, by far, was found by one of Raper's lab technicians, a bacteriologist named Mary Hunt, who had been charged with the task of visiting Peoria's markets in search of moldy fruit and vegetables. In 1943, she hit the jackpot: a cantaloupe infested with a mold so powerful that it would, by the end of the 1940s, be the ancestral source for virtually all of the world's penicillin.

The Northern Lab team, including Andrew Moyer *(left side, fifth from left)* **and Robert Coghill** *(back table, fourth from left)*

At roughly the same time he sent Mary Hunt on her tour of Peoria's fruit stands, Raper had charged another of his subordinates, a microbiologist and mycologist named Andrew Moyer, with finding a better soil: a superior growth medium for the fungus, one that improved the best speed that Heatley's bedpans had been able to achieve using Czapek-Dox and brewer's yeast. Serendipitously, he had access to an extremely promising replacement. The Northern Lab had been established explicitly to investigate "industrial uses for the surplus agricultural commodities." In practice, this meant searching for some commercially valuable way of exploiting the waste products left after the American corn harvest—in 1940, more than 56 million metric tons, much of which was turned into corn flakes, animal feed, sweeteners, and a dozen other commodities. The most significant of these, corn steep liquor, was what was left behind after extracting cornstarch. In weeks, Moyer and Heatley, working together, discovered that corn steep liquor plus sugar increased penicillin production significantly. Actually, more than significantly—a thousandfold. This is not a misprint. Earlier that year, the Dunn team had defined what became known as the "Oxford unit," the quantity of penicillin that, when dissolved in 1 cc of water, inhibited a standard measure of bacterial growth.* The new growth medium was able to improve production from 2 Oxford units per cc of broth to 2,000.

This left the harvesting problem: fermentation itself, which was a challenge not just of biology, but geometry. The metabolic process by which sugars are converted to acids, gases, and alcohol had many forms, as had been known even before Pasteur, but the *Penicillium* mold had, thus far, fermented only on the surface of a growth medium, usually agar (this is why flat Petri dishes were and remain so common in biology experiments). Since surface fermentation meant only two dimensions were available for the growth of the compound, expanding the harvestable quantities of the mold seemed to require *very* large surfaces—Heatley's bedpans blown up to the size of basketball courts.

It was Robert Coghill, the chief of the Fermentation Division at the Northern Lab, who first proposed using the same sort of deep fermentation used in brewing beer for growing penicillin.

Deep fermentation wasn't a completely novel idea. A German-speaking Czech chemist named Konrad Bernhauer had published dozens of papers on the subject from 1920 onward.† Even earlier, as far back as the First World War, Pfizer had been investigating deep (or, as it was then known, "submerged") fermentation in order to improve yields of what would become their core product: citric acid, the age-old flavoring agent and preservative.

The only way to produce citric acid had been extracting it from fruit, particularly lemons, until the German chemist Carl Wehmer had shown that it was produced by molds as well, specifically *Penicillium*. However, making citric acid by extracting it from a mold was just as

---

* By 1944, this would be adopted as a "unit" of penicillin, the amount of the drug that, dissolved in 50 milliliters of broth, inhibited growth in a colony of *staphylococcus*, which was effectively the specific activity contained in .6 micrograms of the International Penicillin Standard, essentially the original Oxford unit.

† Because Bernhauer was an early and fanatic member of the Nazi Party, much of his pioneering work doesn't appear in the standard histories of fermentation.

likely to produce oxalic acid, which was both unwanted and dangerous. In 1917, Pfizer's brilliant research chemist James Currie* discovered that *Aspergillus niger* was a factory for citric acid—feed it sugar, harvest the acid—and started the program the company christened SUCIAC, Sugar Under Conversion to Citric Acid. Enough citric acid, in fact, that, by 1929, Pfizer was selling more than $4.5 million of it.

Such fermentation was aerobic, with *A. niger* exposed to air using Heatley-like shallow trays. By 1931, though, Pfizer's chemists had graduated, if that's the right word, to production of citric acid in relatively small flasks, about 1 liter in volume, in which a powerful stirrer kept the fluid aerated, a process for which they applied for a patent.

At the time Moyer and Coghill started their researches, the technique had been used only for citric acid. Why not penicillin? As Pasteur had been the first to notice, all forms of fermentation are largely similar. In theory, therefore, the same industrial processes used to manufacture citric acid—or, for that matter, beer—could be enlisted for the new miracle drug. The differences weren't trivial: Beer and citric acid could be fermented in relatively unclean environments, but the air used by fungal cells to produce penicillin needed to be sterile in order to avoid introducing dangerous impurities; the temperature within the hypothetical vat needed to be kept constant; and the process required some way of keeping the whole mess mixed so that each liter had the same amount of growth medium and mold as every other. But they weren't insuperable. As early as 1937, the USDA's By-Products Lab at Ames, Iowa (a predecessor of the Northern Lab), had designed an aluminum rotary fermenter. By the fall of 1941, the Peoria team had a demonstration vat—a drum with a washing machine–like agitator, and an injector through which sterile air could be constantly introduced to

---

* Currie, like Charles Thom, the mycologist who directed Florey and Heatley to Peoria, was a graduate of the Department of Agriculture's Research Division. In fact, the two had been collaborators on the department's attempt to produce an American version of Roquefort cheese, which required both extensive knowledge of fermentation, and, coincidentally enough, the variant of the penicillin mold known, for obvious reasons, as *Penicillium roqueforti*.

the soupy contents. Rotary drums like it would be manufacturing penicillin in industrial quantities for the next five years.

At the same time that the Peoria scientists and engineers were cultivating ever more powerful and pure strains of penicillin, the new drug's potential was exceeding even the hopes of its most passionate advocates, particularly in Britain. Henry Dawson, at Columbia University College of Physicians and Surgeons, injected two patients with a crude filtrate of penicillin broth, though at such a low concentration that it was both safe and ineffective. Between June and August 1941, five more staph-infected patients in Britain—more volunteers from Oxford's Radcliffe Infirmary—were treated with it; three of them were children, precisely because the quantity of the drug was so limited, and a child could be expected to respond to a lower dose. On August 16, the *Lancet* featured another article from the Dunn team that reported "that in all these cases a favourable therapeutic response was obtained . . . ," though in one—the tragic case of four-and-a-half-year-old John Cox—the penicillin cured the staph infection that had caused septicemia in his sinus orbits, lungs, and liver, but could do nothing about the ruptured spinal aneurysm that killed him two weeks after he started antibiotic treatment.

In September 1941, Florey, having completed his proselytizing in the United States, returned home to England to continue his own investigations, leaving Heatley behind to work at the Northern Lab. He was, in consequence, not on hand when the tiny vial of brown penicillin powder that saved the life of the septicemic Anne Miller in March 1942 had, in some sense, marked not just the birth of the antibiotic age, but also the end of European, and especially British, preeminence in the field. The following month, Florey opened a package from the United States expecting to find his long-awaited kilogram of penicillin. It contained only five grams; less, in fact, than the amount that Merck had rushed to New Haven Hospital to save Mrs. Miller's life. The Americans were keeping everything they could produce for their own research; despite Florey's tireless campaigning during the summer and fall of 1941, British research on penicillin still relied on what Britain could produce at home.

While Heatley and Florey were traveling across North America drumming up support for their original request—that elusive kilogram of penicillin needed back in Oxford for clinical testing—the Dunn School had remained the most important antibiotic research center in the world, though still reliably dysfunctional. Chain had learned that his colleagues were leaving for the United States only when he noticed their readied luggage. Not only had he not been invited; he hadn't even been informed. He was furious.

The rationale—that the purpose of the trip was to promote industrial manufacture of the drug, which was Heatley's special province, not to discern its structure, which was Chain's—was strong, but unpersuasive. Though Florey understood, earlier than most, that penicillin would never realize its potential either as a therapy or as a scientific breakthrough until its structure was completely understood, he had misunderstood his émigré colleague. Chain was even more avid for recognition than Florey, and feared that he would be cheated out of "his" Nobel Prize—yes; he was already dreaming of a call from Stockholm—unless he promoted his involvement in the discovery.

He wasn't just sensitive to slights concerning his scientific importance. He was equally sensitive (probably more understandably) about his status as a Jewish émigré. Though Chain could not yet know the fate of the mother and sister he had left behind—both would die in 1942, likely in the Theresienstadt concentration camp—he was well aware that the world had rarely been so dangerous for Jews. In his own words, "One could not trust any undertaking given by Florey. I gave in in the end . . . to avoid any action which could provoke latent anti-Semitism, which was very widespread. I had to bear in mind the fact that the Jewish community would not be very pleased if controversies of any kind with anti-Semitic overtones came into the open. . . . I have always considered, and still do, that Florey's behaviour to me in the years 1941 until October 1948, when I left Oxford . . . was unpardonably bad."

During the first months of 1942, Chain and Abraham finally perfected the chemical technique required to produce a stable penicillin salt. The tiny quantities of the drug still being produced by Heatley's jury-rigged factories were producing better and better results on patients from the Radcliffe Infirmary, and, at the instigation of Winston Churchill, Britain's drug companies were finally stepping up to the challenge. By the fall of 1942, Kemball, Bishop was sending 150 gallons of broth to Oxford every ten days, which allowed the Dunn—by far the most productive penicillin "factory" in Britain—to supply sufficient research material to Ethel Florey during 1942 and well into 1943.

Nonetheless, the other Dr. Florey recognized that the center of gravity in penicillin production was shifting irrevocably to the United States, the only place with sufficient penicillin for clinical trials. Florey's original research program had consisted of three key objectives: finding more potent penicillin strains, increasing manufacturing capacity, and unlocking the compound's chemical structure. Two of them were rapidly outgrowing England. Peoria was well on its way to identifying the most promising strains of *Penicillium*, and—as will be seen—American industry was about to be enlisted in manufacturing the precious stuff in quantity. The third goal, however, remained on the Dunn School's research agenda: discovering penicillin's elemental components, its structure, and its mechanism of action—how it worked.

As can't be repeated too often, it's a great deal easier to discover a compound's ingredients than to understand how they fit together. Fairly rudimentary tools, for example, can identify the presence of sugar, flour, butter, and eggs in a cake; they will tell next to nothing about how the chains of proteins, carbohydrates, and fats link together. And they tell nothing at all about oven temperature or baking time. So, too, with the chemistry of penicillin. The ultimate goal of all the researchers was a method of producing penicillin in a factory, rather than growing it in fermentation tanks, to ensure both consistent quality and increased quantity. It was the same way that Ehrlich created Prontosil: analysis, then synthesis.

This is not to imply that analysis of penicillin was a trivial matter. Teams on both sides of the Atlantic struggled for more than a year to establish the type and number of elements that composed the penicillin molecule. As early as 1940, the Dunn team had demonstrated that,

Credit: Getty Images/Howard Clements

Dorothy Crowfoot Hodgkin, 1910–1994

When Dorothy Mary Crowfoot arrived at Oxford's Somerville College in 1928—she didn't marry Thomas Hodgkin until 1937—the
science of X-ray crystallography was barely a decade old, though the
principle of diffraction had been known for centuries, since the Scottish mathematician James Gregory noticed that rays of sunlight split
into a spectrum of colors when passing through a bird's wing; light-
wave spectroscopes had been used to analyze chemical elements
ever since. X-rays, on the other hand, which had been discovered by
the German physicist Wilhelm Roentgen in 1895, travel in waves that
are thousands of times shorter than light waves. This meant that the
prisms and other tools used to split light waves were too coarse a filter,
about as useful for diffracting X-rays as a fishing net was for catching
gnats. In 1912, though, another German physicist, Max von Laue, proposed that the tightly packed and repetitive atoms found in crystals—
copper sulfate, for example—could do the job. Three years later, the
father and son British physicists William Henry Bragg and Lawrence

Bragg described the math that could decode the pattern of myriad dots recorded on a photographic plate after scattering X-rays through a crystal: If the X-ray wavelength and the intensity of the image were known, the molecular structure of the crystal could then be calculated. Table salt, for example, as the Braggs showed, has a *formula* of one atom of sodium plus one of chloride: NaCl. But the X-ray diffraction pattern shows that it is organized in alternating cubes, an atom of sodium surrounded by six chlorine atoms, next to a single chlorine surrounded by six sodium, repeating one after the other.

Dorothy Crowfoot was barely fifteen, and still living in Khartoum— her father, John Winter Crowfoot, was employed in the Egyptian Education Service of Britain's Colonial Office—when Dr. A. F. Joseph, a chemist working for Wellcome Laboratories, gave her a book by Sir William Bragg on X-ray diffraction, and so kindled her fascination with crystals. At Oxford, the fascination turned into a passion. Under the tutelage of Robert Robinson, a future chemistry Nobelist and the head of the Dyson Perrins lab, and especially J. D. Bernal, she mastered the technique of using X-ray crystallography to map the internal structure of organic molecules several orders of magnitude more complex than Bragg's table salt.* The list included cholesterol, testosterone, progesterone, pepsin, and a dozen more, most particularly insulin, whose structure Hodgkin began investigating in 1934, and finally resolved thirty-five years later, in 1969.

By 1940, Hodgkin had begun a relationship even more fruitful and long-lasting than the one with Bernal, accepting her first grant, in the amount of £1,000, from Warren Weaver of the Rockefeller Foundation (though, Weaver, whose eye for talent had already found Howard Florey and Ernst Chain, described her request for a grant "for research on proteins and viruses, with the object of carrying out more general and fundamental investigations . . ." as an "uncertain

---

* Bernal, another of the key figures in the development of X-ray crystallography, was not only Hodgkin's mentor, but her lover. Both were devoted Socialists as well, and lifelong supporters of the Soviet Union; Bernal so much so as to become Britain's most passionate advocate for the bogus genetic theories of Trofim Lysenko.

proposal at the present moment"). It was the beginning of a nearly thirty-year relationship with the Rockefeller Foundation, one of the longest in the foundation's history.

Also in 1940, her extraordinary talents—and her team at Oxford's Laboratory of Chemical Crystallography—were recruited to the penicillin project. Even before the publication of the August 1940 article in the *Lancet,* she had been given a heads-up about the miraculous discoveries going on at the Dunn School, when she encountered Ernst Chain on a walk along Oxford's South Parks Road, just after the first mice had been successfully treated with penicillin. Chain promised, "Some day we will have crystals for you." It took more than a year for the promise to be redeemed, but in November 1941 Hodgkin wrote to her husband, "I've just come back from visiting Chain and now it's 10:30 P.M. I'm feeling disgustingly cheerful as a result of my visit. . . . Apparently [penicillin] hasn't yet been crystallized after all. . . . Chain seemed quite keen to let me have some stuff"—the degradation products—"and I'd simply love to try."

She quickly found that the degradation products were extremely difficult to work with: tiny crystals that were, in Hodgkin's words, "immersed in a gummy fluid [and] extremely hygroscopic [i.e., attracting water molecules from the environment], so that they are practically impossible to leave out in the air for more than a few minutes or so." They did, however, reveal one of penicillin's secrets: Sulfur was an essential element of the compound, which seemed to be a molecule containing nine atoms of carbon, eleven of hydrogen, four oxygen, two nitrogen . . . and one atom of sulfur.[*]

But how were the atomic bricks arranged into a structure that was able to stop infectious pathogens cold? X-ray crystallography was a powerful tool for investigating such questions, but not an easy one. Creating the picture formed by the X-rays scattered by penicillamine

---

[*] The first recipe for penicillin was in the form of a sodium salt. As a barium salt, the chemical formula was $(C_{14} H_{19} O_4 N_2 S)_2$ Ba, and was named penicillin F. A slightly different version was known as penicillin G.

crystals onto a photographic plate was only the first step, and as likely to be confusing as to be revealing.

The reason is that the intensity of the dots captured on a photographic plate measures the amplitude of the wave of the X-ray as it bounced off the faces of the compound's crystalline lattice, from which a complete picture of the crystal could, in principle, be derived. More electron density, higher amplitude. But a dot that made a dense, high-intensity impression on the plate could just as easily be the result of two waves that were both arriving at the same frequency—the same number of waves per second—and amplitude. When those kind of signals arrived "in phase" with one another, they reinforced the density. In the same way, a low-density image could be produced by two waves that were out of phase and canceled one another out.

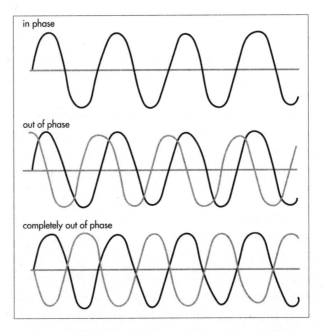

Phases: In the top graphic, two waves are in phase—that is, completely overlapping. In the middle, the waves are out of phase, and, in the bottom, completely out of phase.

Solving the potential confusion demanded a subtle and tricky mathematical technique, one that had originally been developed in the early nineteenth century to describe how waves of heat travel in three dimensions over time. As an example, if a match is held under the center of a coin, the coin's center will heat up; after the match is extinguished, the heat will dissipate in an ordered way, with the center of the coin getting colder, and the edges warming, until an equilibrium is reached. This phenomenon, however, is an extremely complex business, which is why, in the early nineteenth century, the French mathematician Jean-Baptiste Joseph Fourier helped to develop a technique that turns complicated wave functions like these into a collection of the simple sine or cosine waves that are taught to high school trigonometry students.

Unfortunately, while X-ray amplitudes are relatively easy to chart with a simple molecule—remember the Braggs' table salt—they rapidly become more complicated as molecules get larger. The electron density photographs start to resemble three-dimensional weather maps, complete with eddies, currents, and whirlpools. Diagramming a molecule the size of penicillin—essential for a full understanding of its structure—and calculating which waves were in or out of phase, would take Hodgkin more than two years.

In the meantime, on both sides of the Atlantic, progress was accelerating in the race to produce penicillin in useful quantities: more animal trials, increasingly sophisticated production methods, and, of course, more precise maps of the mysterious compound. Most of the key work was invisible to the public at large; though, on the heels of the Dunn team's August 1941 *Lancet* article, the September 15, 1941, issue of *Time* magazine reported:

> *A marvelous mold that saves lives when sulfa drugs fail was described in the British* Lancet *last month by Professor Howard Walter Florey and colleagues of Oxford. The healing principle, called penicillin, is extracted from the velvety-green* Penicillium notatum, *a relative of the cheese mold. Although it does not kill germs, the mold stops the growth of* streptococci

and staphylococci *with a power "as great or greater than that of the most powerful antiseptics known."*

The article didn't really generate much in the way of follow-up. Neither, implausibly, did the cure of Anne Miller in March 1942. From the standpoint of scientists like Heatley, Coghill, Florey, and even Chain, that was as it should be. With barely enough penicillin being produced for the minimum number of human trials, the last thing anyone needed was to raise the hopes of every patient in the English-speaking world. But, as the tide of war against the Axis seemed to slowly shift in favor of the Allied armies, the public, in the United States and Britain particularly, was primed to embrace a life-saving miracle, the world's first antibiotic drug.

And, of course, the hero who discovered it.

At the end of August 1942, the *Times* of London published a brief editorial promoting a greater degree of public investment in penicillin development. Either out of ignorance or delicacy, the editorial declined to lionize (or attack) any scientist or public official by name.

It's impossible to know what reaction the editorial might have generated on its own. However, on August 31, the *Times* published the following letter:

Sir,—In the leading article on penicillin in your issue yesterday you refrained from putting the laurel wreath for this discovery around anyone's brow. I would, with your permission, supplement your article by pointing out that, on the principle of palma qui meruit ferat,* *it should be decreed to Professor Alexander Fleming of this research laboratory. For he is the discoverer of penicillin and was the author also of the original suggestion that this substance might prove to have important applications in medicine.*

---

* *Palma qui meruit ferat* translates as "Whoever earned the palm should wear (or bear) it," and was the motto of Lord Horatio Nelson. Also, and completely coincidentally, it is the motto of the University of Southern California.

As the careful reader might have guessed from the phrase, "this research laboratory," the letter was signed:

Almroth E. Wright
Inoculation Department, St. Mary's Hospital

Fleming's boss and mentor, who remained one of Britain's most famous physicians, had spoken. Within hours, Fleet Street emptied out and headed to Paddington to find Fleming and Wright. An interview with Fleming appeared in the *Evening Standard* of August 31. On September 1, half a dozen other papers reported on Fleming's discovery. Fleming was named the "Man of the Week" by the *News Chronicle*. Some stories actually had the lab at St. Mary's producing the samples used at Oxford. In the *Daily Mail*, Fleming even gave the very strong impression that St. Mary's, which had last investigated the properties of penicillin seven years before, was the place where a breakthrough could be expected, saying, "the production of the drug is very complicated and the difficulties are great but they are being overcome."

The obvious implication—that St. Mary's was the center of penicillin research—didn't go uncommented. English organic chemist and Nobel laureate Sir Robert Robinson had already replied to Almroth Wright with *his* letter to the *Times*, which appeared on September 1 and which read, in part, that if Fleming should wear the laurel wreath, then "a bouquet, at least, and a handsome one, should be presented to Professor H. W. Florey . . . he, and his team of collaborators, assisted by the Medical Research Council, have shown that penicillin is a practical proposition."

Florey was glad of Robinson's support, but angry nonetheless. On December 11, 1942, he wrote to Henry Dale, president of the Royal Society, "I have now quite good evidence, from the director-general of the BBC in fact . . . that Fleming is doing his best to see the whole subject is presented as having been foreseen and worked out by Fleming and that we in this department just did a few final flourishes."

Public confusion about the discovery of penicillin goes on. Recreations in a BBC documentary from the 1970s include a dramatized Fleming preparing the compound to treat Albert Alexander, and even got the year wrong. W. Howard Hughes's biography of Fleming, *Alexander Fleming and Penicillin,* claimed that St. Mary's technicians had been making penicillin every week since its discovery in 1928. The durability of the resentments expressed by the Dunn School team are, in their way, enlightening: a window onto the motivations of research scientists, where careers can literally rise and fall over a few days difference in announcing even a minor discovery to other professionals. Scientists are human beings, after all, and their reasons for spending thousands of hours in poorly ventilated laboratories for poverty wages are complicated: the thrill of solving difficult puzzles; the joy of exercising their talents cultivated over a lifetime; and, of course, the pride and adulation that are the reward of first discovery.

There are practical consequences to primacy of discovery. As Almroth Wright no doubt realized, the institution that could claim penicillin for its own would be a repository of both public acclaim, and a greater share of the financial resources of philanthropies and government agencies. The ways in which a legitimate claim to discovery profited the discoverers of penicillin—like the discovery of Salvarsan or Prontosil before—remain huge, even without calculating the benefits to society at large.

But those ways pale in comparison to the very tangible profits that would be generated by the discoveries that followed.

FIVE

"To See the Problem Clearly"

The battles over credit for the discovery of penicillin were still a year in the future when Howard Florey left Peoria in the summer of 1941. He left Heatley behind to work with Moyer and the rest of the Northern Lab team, while he transformed himself into North America's most prominent penicillin evangelist. Florey had already spread the gospel to the National Research Council in New Haven, and to the USDA in both Washington and Peoria. On August 7, he arrived in Philadelphia for what would be his most consequential meeting of all, with Alfred Newton Richards of the University of Pennsylvania. Florey had met and worked with Richards as a junior researcher during his first Rockefeller Foundation–funded trip to the United States in 1925; he now had the most exciting discovery in all of medicine, and A. N. Richards was precisely the man who could translate that excitement into action on a national scale. The previous year, he had become chairman of the Committee on Medical Research for the U.S. government's newly created Office of Scientific Research and Development. This made him, to Florey, the most important person in the country.

By some measures, in fact, the OSRD would be the most important strategic asset of the United States in the coming war. When Florey and Richards met, it was, technically, only a week old, the creation of President Roosevelt's Executive Order 8807, which had established it to guarantee "adequate provision for research on scientific and medical problems relating to the national defense," though it had replaced an earlier version, the National Defense Research Committee, created the previous year. In both incarnations, it was run by an electrical engineer named Vannevar Bush. When Roosevelt tapped him to lead the country's defense-related research, Bush had already made his name as a successful inventor of some of the basic components of what would become both digital and analog computers, as dean of MIT's School of Engineering, president of the Carnegie Institution, and the founder of the company that was then known as the American Appliance Company (but would evolve into the electronics giant Raytheon). As the first and only head of the National Defense Research Committee, he was, in effect, the first presidential science advisor; as head of the OSRD, he gave strategic direction to the most consequential wartime research projects in history, most notably, the Manhattan Project, which would usher the world into the atomic age.

It's a close call, though, whether the long-term consequences of nuclear power would be more significant than those of the antibiotic revolution that was ignited by the OSRD's Committee on Medical Research in 1941. The committee was explicitly *not* set up to initiate original research, but rather to oversee existing programs and to set up protocols for funding through its six divisions: medicine, surgery, aviation medicine, physiology, chemistry, and malaria. Nonetheless, Florey emerged from his August 7 meeting with Richards, agreement in hand guaranteeing the CMR would recommend a government grant for the production of penicillin.

Medical research paid for, and managed by, federal agencies wasn't a completely novel concept, even then. Well before the 1940s, the U.S. government had directly sponsored significant research on disease treatment and prevention; in 1887, Dr. Joseph J. Kinyoun of the Marine Hospital Service established a bacteriological laboratory in the Marine Hospital in Staten Island. In 1891, it was moved to Washington, DC, and renamed the Hygienic Laboratory. In 1902, Congress passed the Biologics Control Act largely to regulate the vaccines sold across state lines, but the same act also authorized the Hygienic Laboratory to test and improve such products as vaccines and sera—and added divisions in chemistry, pharmacology, and zoology. By 1912, the Marine Hospital Service had been transformed into the U.S. Public Health Service; in 1930, the Hygienic Laboratory had been renamed the National Institutes of Health, and seven years later moved to Bethesda, Maryland.

But previous investments in research had funded institutions administered by the federal government: the USDA's laboratories, or—a very different kind of research—Massachusetts's Springfield Armory. No governmental-business alliance like the penicillin project had ever been contemplated. On October 8, 1941, Richards and Bush called a conference to be held in Washington. Among the invitees were a number of OSRD department heads: Lewis H. Weed, the vice chairman of Richards's Committee on Medical Research; William Mansfield Clark, the chairman of the Division of Chemistry; and Charles Thom from the Department of Agriculture. More remarkably, the meeting, at which Bush presided, also included George A. Harrop of Squibb's Institute for Medical Research, Jasper Kane of Pfizer, Yellapragada Subbarao of Lederle, and Randolph Major, the research director of Merck & Co.*

The ad hoc committee would meet again on December 17, in New York. This time, the drug companies were represented not just by their heads of research, but by their presidents. George Merck, the president of his family company, presciently observed, "If these results could be confirmed . . . it was possible to produce the kilo of material for Florey. . . ." But by that time, ten days after Pearl Harbor, the nation had considerably more ambitious goals. Richards was no longer looking for a kilo of material from a single interested corporation. Now, "every possible means of combating infection in battle casualties [must] be explored." Robert Coghill of the Northern Lab, who was also present at the meeting, would later confirm: "A new pharmaceutical industry was born."

With the birth of that new industry came a raft of other issues. Despite the patriotic enthusiasm with which the companies embarked on the first stages of penicillin manufacture, they were still, after all,

---

* Years later, in 1957, Bush was named chair of Merck's Board of Directors. It was a largely ceremonial job, but not to Bush. A year after his arrival, he was writing to George Merck scolding him that his corporate organization was "atrocious" and Merck himself far too easygoing.

commercial entities. And, although the commercial potential of penicillin was more than alluring to them, they soon realized that the lasting value of this project was knowledge. At the very least, some mechanism was needed to resolve disputes about the ownership of that knowledge: a patent.

The historian of science Derek de Solla Price is generally credited with the observation that patents are to technology what scholarly papers are to science: the key method of allocating credit and disseminating knowledge. But while a system of patents had been a much-lauded feature of American society virtually since the country's founding—Article I, Section 8 of the U.S. Constitution explicitly empowered the federal government to provide limited patents "to promote the progress of science and useful arts"—they had never been entirely free from controversy. Thomas Jefferson was, at least initially, hostile to the very notion of patenting ideas, writing, "If nature has made any one thing less susceptible than all others of exclusive property, it is the action of the thinking power called an idea. . . . Inventions cannot, in nature, be a subject of property." As they were applied to medical innovations, patents were even more controversial. While mechanical inventors were unapologetically commercial in their goals, medicine—in theory, at least—was subject to a higher law, one that required that its work be performed for the greater good. Patenting of medicines, in consequence, had been forbidden in France from the time of Napoleon forward, and in Germany the nation's first patent law, which was passed shortly after the modern state appeared in the 1880s, prohibited them for decades.

American medical researchers had a very specific objection to medical patents, dating from 1923, when Harry Steenbock, a professor of biochemistry at the University of Wisconsin, discovered that exposing the sterols in fatty foods like milk to ultraviolet light enriched them with vitamin D, and therefore made milk into a defense against the then-widespread deficiency disease, rickets.* Although

---

* In 1921, three-quarters of the children in New York City showed signs of rickets: bow legs, joint pain in the extremities, and deterioration of teeth and bones.

Steenbock, who had patented the process in his own name, reportedly turned down a million dollars from the Quaker Oats Company for its use, he did transfer the patent rights to a newly established nonprofit organization, the Wisconsin Alumni Research Foundation (WARF). By 1940, WARF was charging milk manufacturers for the use of the technology, and had pocketed royalties well in excess of $7.5 million (the number today is well in excess of $1 billion).*

To the surprise of exactly no one, this was a source of resentment, followed by reaction. Dozens of research institutions, including Harvard, the University of Pennsylvania, Johns Hopkins University, and the California Institute of Technology, either forbade or severely restricted the ability of researchers to seek patents. By 1937, the American Chemical Society was hosting a conference entitled "Are Patents on Medicinal Discoveries and on Foods in the Public Interest?"

On the other hand—when it comes to intellectual property, there is *always* at least one other hand—Ernst Chain had proposed patenting the Dunn School compound from the time it had been successfully extracted. By March 1941, he was lobbying Dr. J. W. Trevan, then the director of the Wellcome Research Laboratories, to support a patent application. He had considerable support from his boss; Florey shared Chain's eagerness to secure a patent, less as a way to enrich the Dunn researchers than to enlist an entity like the Medical Research Council to whom such patents could be assigned.

Unfortunately, though, the Medical Research Council wanted nothing to do with it. Its director, Edward Mellanby, the discoverer of vitamin D, had led the fight to invalidate Steenbock's patent in Britain, and regarded patents with the sort of distaste that wealthy aristocrats generally hold for parvenu tradesmen. It would be a longstanding point of contention between Mellanby and Chain, who saw the development of antibiotics as "a whole tremendous virgin field [in which] we were the leaders and would remain so *if we got enough*

---

* Not coincidentally, the patent was also used to prevent manufacturers of margarine from enriching their product by irradiation, which was a matter of some significance to the most important agricultural industry in Wisconsin.

money" [emphasis added]. He later said that it was unethical *"not* to take out patents protecting the people in this country against exploitation by foreign commercial organizations. . . ."

In the event, he didn't persuade Mellanby, who rebuked Chain for his stubbornness about patenting, telling him that if he "persisted in his 'money grubbing' he would have no scientific future in Britain." In this, he was allied with the Rockefeller Foundation, which had a policy that discoveries generated on their dime ought to be free of patent . . . and they had been bankrolling Florey and the Dunn since 1939.

But by the end of 1941, both the Rockefeller Foundation and the Medical Research Council had lost control over the penicillin project, which was now largely controlled by the OSRD, American pharmaceutical companies, and the United States Department of Agriculture, all of whom were enthusiastic about patents, less as a financial incentive than as a way of managing the diffusion of new knowledge in a systematic way. The enthusiasm had a strong historical foundation: Henry Leavitt Ellsworth—the "Father of the USDA"—had established the ancestor of the USDA as the Patent Office's Agriculture Division in 1839.

Small surprise, then, that in the fall of 1941, all the parties working at the Northern Lab, including Norman Heatley, signed a letter of agreement that assigned any subsequent patents to the United States Secretary of Agriculture. No doubt it seemed noncontroversial at the time.

Heatley stayed at Peoria until June 1942, working with his American counterpart, the mycologist Andrew Moyer. Despite Heatley's naturally genial disposition (and his practice dealing with the extraordinarily difficult Ernst Chain), the relationship was the opposite of amicable. In Heatley's recollection, Moyer was a loud and obnoxious isolationist convinced Britain was pulling the United States into a mistaken war, one that would inevitably demand "gutters overrunning with [presumably American] blood."

Even more troubling in the long term, Moyer applied for and received a patent not only for the deep fermentation methods that he had developed jointly with Heatley—the patent itself reads: "a new and useful method for producing penicillin by the cultivation of molds, whereby the yield of penicillin is substantially increased above that previously obtained"—but for using the corn steep liquor. The names on the patent were Robert Coghill and Andrew Moyer: no mention of anyone from the Dunn School. In fact, though Heatley and Moyer had collaborated on a paper summarizing their joint research at Peoria, Moyer never submitted it for publication. It appears neither in his personal bibliography nor the patent application itself. And, while the underlying patent was assigned to the U.S. Secretary of Agriculture, allowing use without payment by all American users, it said nothing about the ability of the patent holders to sue for compensation in countries that were signatories to patent treaties with the United States. One such country was the United Kingdom, which allowed Moyer to secure British patent rights for his method of production.

The members of the Oxford team were furious. Chain, in particular, spent the rest of his life feeling wronged over the patent issue. He had some justice on his side, but not much. The fact was that patenting penicillin itself was highly problematic, since the original discovery, of course, dated from 1928. The elapsed time meant that a patented *process for manufacturing it* was far more valuable than the substance itself. Chain's grievances were emotional, not legal. The true innovations in manufacturing the substance had been developed in Peoria, not Oxford. The key Coghill-Moyer patent, for example, was for adding phenylacetic acid to the penicillin broth, which increased yields by two-thirds. This isn't to excuse Moyer's actions, which were duplicitous, at best. But since the patents in question were for the procedures developed in 1941 at the Northern Lab, Heatley was the one with a reason for feeling wronged, not Chain.

If Heatley had any complaints about the loss of credit, he had little time to share them. Even before the patent had been granted, Florey sent Heatley a cable reading, "WHY NOT GO MERCK SIX MONTHS IF THEY WILL PAY YOU. MORE USEFUL THAN COMING BACK HERE."

When Norman Heatley left Peoria for Rahway, New Jersey, in 1942, Merck was the oldest continuously operating drug firm in the world; though as we have seen, this may be damning with faint praise. The company had been producing and selling medicinal compounds since 1668, when Friedrich Jacob Merck acquired the Engel-Apotheke, an apothecary in the Landgraviate of Hesse-Darmstadt, one of the dizzying number of German-speaking principalities that made up the seventeenth-century Holy Roman Empire.

The business of selling medicine to consumers and physicians remained for nearly two centuries the Engel-Apotheke's sole business, though one with limited potential before Pasteur and Koch established the microbial causes of infectious disease. In 1816, one of Friedrich Jacob's descendants, Heinrich Emanuel, became proprietor of the family business, changed the name to E. Merck, and put the company into a different business altogether. An ambitious and, for his day, scientific druggist, Emanuel Merck sensed that the chemicals known as alkaloids were components of a large number of powerful plant-based extracts—examples include belladonna and caffeine—that could be purified and standardized. The most powerful of these extracts was the one derived from *P. somniferum*, the opium poppy, and in 1827 Merck bought from a Prussian apothecary named Friedrich Serturner the process by which opium could be transformed into a compound he called morphine, taking the name from the god of dreams in Ovid's *Metamorphoses*.

For a century, morphine would be the most profitable and popular medicinal product produced by Merck, but its real importance was giving the still-small company expertise in chemical manufacturing. The same techniques used for making consistent doses of morphine out of decidedly impure opium allowed Merck entrée into the business of manufacturing so-called fine chemicals: small-batch, high-value, and very pure compounds. His timing was excellent, as the giant chemical companies that emerged after William Henry Perkin's discovery of the first aniline dyes became Merck's largest customers, and a valuable source of knowledge about the most advanced nineteenth-century industrial chemistry. In 1889, the company's first publishing venture, *Merck's Index*, became the ultimate source of chemical information for a generation of scientists and engineers.

Merck remained a maker of medicines as well. In 1887, the company opened its first office in the United States, as a marketing and sales department for the German parent, still known primarily for morphine; and, in 1891, as a subsidiary: Merck & Co., run by Georg Merck. The following year, Georg anglicized his name by adding an *e* to it, and welcomed his first child: George W. Merck.

For the next twenty-five years, the company prospered. In 1900, it acquired 120 acres of swampland in Rahway, New Jersey, where it built a factory to produce bismuth—the active ingredient in Pepto-Bismol, invented as an antidiarrheal in 1901—cocaine, and morphine. Next door, another Merck subsidiary known as Rahway Coal Tar Products manufactured, among other things, carbolic acid, the antiseptic discovered by Joseph Lister four decades before. By 1917, Merck & Co. was recording sales of $8 million annually—roughly $96 million today—solid, but not earthshaking, considering that same year General Electric sales were nearly $200 million, and U.S. Steel $400 million.

Of course, 1917 is the year that marked the entry of the United States into the First World War, with serious consequences for U.S. subsidiaries of German companies. Because Germany, particularly the I. G. Farben cartel, dominated manufacture of virtually all medicinal chemicals, supplies of critical compounds were at risk. The world's entire supply of atropine, for example, an extract of belladonna that was one of the key medications in treating heart problems, was in the control of Germany and her allies. On April 17, 1917, the Medical Section of the Advisory Commission of the U.S. Council of National Defense convened a meeting attended by representatives of more than 250 companies, including Merck, to find alternatives.

Meanwhile, anti-German hysteria led to a dozen new laws intended to limit the activity of "enemy agents," such as German-owned corporations. The Trading with the Enemy Act, which went into effect in October 1917, was the most significant. One of its clauses provided for the selection of an Alien Property Custodian, a federal judge named A. Mitchell Palmer who immediately required that the German-owned chemical companies be "Americanized." True to his word, he confiscated 80 percent of Merck, the eight thousand shares owned by the parent company in Hesse-Darmstadt—and only held them in trust after a plea from George W. Merck, Sr., who owned the other two thousand shares himself, and didn't want his company sold out from under him.*

At the moment the roof fell in, twenty-nine new researchers—chemists, pharmacists, and chemical engineers—had just been hired. They were just in time to help the company grow dramatically during the war years, which saw the Rahway plant triple in size—and to see George Merck outbid Monsanto and American Aniline to repurchase his company at auction, in 1919, for $3.75 million. Some of those researchers were still there in 1925, when George W. Merck, Jr., took over the firm.

Biographers looking for a successful business leader with a virtually unblemished record for honorable behavior would probably stop searching immediately after coming across George Merck, Jr. Born in New York, he was raised in the comfortable New Jersey suburb of Llewellyn Park where, at least in family legend, he was so friendly with Thomas A. Edison's two sons that he was permitted to haunt the great inventor's lab and workshop. Supposedly, Edison called the Merck heir "Shorty," though he was probably the last to do so. George, as an adult, was an imposing six foot five inches tall, athletic, and charismatic. When he graduated from Harvard as part of the class of 1915—a year early—the war in Europe prevented him from pursuing his earlier plan to attain a graduate degree in chemistry in Germany, and he instead joined the family firm, becoming a vice president by 1918, and, when his father fell ill in 1925, president.

Although it would be reasonable to assume that his rapid promotions were a function of his name rather than his talent, from the start he did his best to correct any misapprehension about his abilities. In 1927, still in serious debt from the purchase from the Alien Property Custodian, George engineered a merger with another fine

---

* Merck wasn't singled out: Bayer's American subsidiary was the largest German company sold, for more than $5.1 million. Cassella Manufacturing and Farbwerke Hoechst were held in trust until the war's end.

chemical manufacturer, the Powers-Weightman-Rosengarten Company of Philadelphia (Adolph Rosengarten, Merck's largest nonfamily stockholder, was ready to retire), and in 1929 made the far more momentous decision to build, on the Rahway property, a research laboratory. At its dedication on April 25, 1933, more than five hundred spectators heard Sir Henry Dale, the future president of England's Royal Society (and the same man who would try and fail to interest British pharmaceutical companies in the penicillin innovations under way at the Dunn School a decade later), give a speech on "Academic and Industrial Research in the Field of Therapeutics." And they heard George Merck assert, "We have faith that in this new laboratory . . . science will be advanced, knowledge increased, and human life will win a greater freedom from suffering and disease."

This corporate gamble on the future of scientific medicine wasn't unique to Merck. The onetime Abbott Alkaloidal Company, renamed Abbott Laboratories in 1914, opened a 53,000-square-foot Chicago research facility in 1938 (interiors by the design genius Raymond Loewy). That same year, in October, the Squibb Institute for Medical Research was dedicated in New Brunswick. In Indianapolis, Eli Lilly opened Lilly Research Laboratories in 1934. A year later, DuPont opened the Haskell Lab of Industrial Toxicology in Newark, Delaware.

But the Merck Institute for Therapeutic Research was the first, and by most measures, the most innovative. In conscious imitation of the academic-industrial conveyor belt that had served the German chemical companies so brilliantly, Merck hired the Viennese pharmacologist Hans Molitor as the institute's first director, and recruited Randolph Major from Princeton to join Alfred Newton Richards—Howard Florey's erstwhile mentor from the University of Pennsylvania—in Rahway.* He invested in more than just personnel; the institute's annual

---

* When Merck approached Richards in 1930, the American Society of Pharmacology and Experimental Therapeutics banned anyone associated with a commercial firm from membership. Richards, who "saw in them no signs of the horns or tail," accepted Merck's offer, and submitted his resignation to the society. Faced with a choice of losing its most respected member, the pharmacological society swallowed its pride,

research budget increased from $146,000 in 1933 to nearly $1 million by the beginning of the 1940s.

If George Merck's goal was that "science will be advanced, knowledge increased," he could scarcely have been disappointed. In 1937, Major persuaded the brilliant Max Tishler—in the words of a colleague, "Max was born with an energy level that was like an avalanche and a brain that was incandescent"—to leave Harvard for Rahway. During the institute's first five years, its researchers published thirty papers in peer-reviewed journals; from 1939 to 1941, the number was closer to fifty. Not included in that number is a nonetheless revealing article that the president himself published in 1935, in the journal *Industrial and Engineering Chemistry*. Entitled "The Chemical Industry and Medicine," it contains the lines, "to do research worthy of the name, to do research which will bring to industry true recognition of its contribution to the advance of knowledge, industry must have at its disposal genuinely creative minds so placed and so protected that the mental powers of thought, study, and imagination can concentrate on problems of great difficulty. For even to see the problems clearly is of itself a major task."

When Heatley arrived in Rahway in 1942, the lab, under Major, was still primarily doing research on vitamins, which then accounted for more than 10 percent of the company's sales. In 1936, Joseph Cline of Merck and Robert Williams of Bell Labs had succeeded in synthesizing vitamin B_1 for the first time, and by 1940 Karl Folkers, late of Yale, had isolated and synthesized both vitamin B_6 and pantothenic acid. Max Tishler had been explicitly recruited to work on vitamins by Randolph Major himself, who told him, "We made up our minds that we're going to specialize in research in the field of vitamins. We're going to isolate every vitamin; we're going to determine their structures . . . synthesize them, and make them available." Tishler didn't waste any time; by 1938 he had discovered a new way to synthesize vitamin B_2 in order to perform an end run around patents

changed its bylaws, and opened its rolls to industrial members . . . possibly one of the most far-reaching achievements of Richards's highly accomplished life.

owned by I. G. Farben and Hoffmann-La Roche, neither of which would license them to Merck.

All that was about to change, not just at Merck, but at other American pharmaceutical firms that had invested millions in research laboratories they were eager to employ on products with more commercial potential than vitamins and antiseptics. In February 1942, Merck signed a research-sharing agreement with E. R. Squibb, the company founded in 1858 by a former U.S. Navy surgeon, Dr. Edward R. Squibb, largely to produce surgical anesthetics. His timing was, in its way, impeccable; the Civil War broke out only two years later, and so increased the nation's annual quota of amputations and other surgical procedures by at least an order of magnitude. Union surgeons carried thousands of the fully stocked wooden medicine chests known as Squibb Panniers from Antietam to Appomattox: a kit of medicinal compounds including anesthetics like ether (which would remain the company's signature product for more than forty years), quinine for malaria, and, of course, whiskey.

By the beginning of 1942, the company had been owned and managed by a former Merck executive, Theodore Weicker,* for more than thirty years; and if there had been any bad blood over his leaving the firm—Weicker had sold his share of Merck's U.S. division in 1903, and, backed by his wealthy industrialist father-in-law, Lowell Palmer, transformed himself from a colleague into a competitor—it had long since been diluted by time, and the national emergency. The Merck-Squibb agreement contemplated joint ownership of any inventions that might appear, not just between the two companies, but including "other firms who have made definite contributions to the solution of the problem."

---

* Weicker was a gifted chemist, but an even better marketer. In 1921, aware of the potential of advertising but wary of the risk of offending physicians, he enlisted Raymond Rubicam, then a copywriter at the Philadelphia agency N. W. Ayer & Son, to square the circle: promote Squibb to the public without insulting the professionals. Rubicam—or possibly Weicker, different accounts give credit to each—came up with a parable, told by "Hakeem the Wise Man," that ended with the moral: "The 'Priceless Ingredient' of every product is the honor and integrity of its maker." Hakeem was discarded soon enough, but the slogan appeared on practically everything that bore the Squibb name for more than twenty-five years.

For most of 1942, the "other firm" that mattered the most was Chas. Pfizer & Co., Inc., of Brooklyn, New York.

Like Merck, Pfizer was a German transplant, though one with a longer history in America. Beginning in the 1840s, a huge wave of German immigrants arrived in the New World, some of them politically motivated—the failed 1848 revolution made thousands into political refugees—but mostly for economic reasons. More than 1.4 million German-speaking immigrants arrived in the United States between 1840 and 1860, and with them came an enormous amount of technical knowledge that its carriers were ready to commercialize. Two of them, Charles Pfizer, an apprentice apothecary, and his cousin Charles Erhart, a confectioner, left the town of Ludwigsburg in the still-independent Kingdom of Württemberg sometime in the 1840s. In 1849, they opened a business office, and later, with a $2,500 loan from Pfizer's father, their first chemical factory, at the corner of Bartlett and Tompkins in the Williamsburg section of Brooklyn.

The cousins' first product combined their chemical training with their skills in candy making, packaging the bitter compound known as santonin—an antiparasitic, or, more precisely, an anthelminthic: a drug used for killing, or at least discouraging, intestinal roundworms—in a toffee-flavored sugar cone. But while flavored santonin remained its most important medical product for more than sixty years, and the company produced disinfectants like iodine, chloroform, and calomel—the highly dubious mercuric compound used to treat both constipation and syphilis—its real business was, like Merck decades later, fine chemicals: tartaric acid (used as the leavening agent in baking powder, and as a flavoring), camphor, and especially citric acid.

It was citric acid that made the company both profitable and perhaps the world's most experienced at the process of deep fermentation. This, in turn, made them ideal to expand on the innovations that Moyer, Heatley, and Coghill had developed in Peoria. Late in 1941, the company had borrowed the same aerated flasks used at its SUCIAC plant to manufacture citric acid and put them to work making penicillin. Yields were still highly variable—from 20 penicillin units per cc to none at all—and tiny, but enough to provide testable quantities to the same team at Columbia University that had requested samples from Ernst Chain back in 1940: Henry Dawson, Karl Meyer, a chemist, and Gladys Hobby, a microbiologist, all part of a group of scientists studying *hemolytic streptococci*. Others were investigating as well. Alexander Hollaender, a researcher at the National Institutes of Health and one of the founders of the field of radiation biology, assembled a team at Cold Spring Harbor lab to find a mutated version of the *Penicillium* mold using radiation. Researchers at the University of Minnesota and the University of Wisconsin were enlisted to bombard them with X-rays; the team at Madison had a pilot plant for deep fermentation, and had been researching irradiation in milk to improve vitamin D content since the 1920s. They were so successful that they ended up developing a new variety of the mold, known as Q-176, which produced more than 2,000 units of penicillin per cc. This improved on Pfizer's yields dramatically, and matched the best number produced by Moyer and Heatley in Peoria. (The original Oxford number, lest we forget, was barely 2 units per cc.)

In less than a year, the seemingly inexhaustible resources of the United States—Department of Agriculture laboratories, university biology departments, and especially the research facilities built by American drug companies in the 1930s—had increased the pace of innovation in penicillin research by an almost unimaginable amount.

All this progress in the manufacture of the drug was well known to the brilliant but underfunded scientists at the Dunn School (to say nothing of St. Mary's), who could only look on in amazement as the United States invested millions in exploiting their discoveries. But except for the occasional "wonder drug" article here or there, general awareness of penicillin's potential for fighting infectious disease was very low in both the United Kingdom and the United States. Through 1942, the OSRD's various subdivisions had authorized only twenty-two specialists in the entire United States to receive the drug at all, and even they were permitted to test it only on a very short list of infections—basically staph, strep, and *pneumococci* infections that didn't respond to the sulfa drugs.

In November 1942, the public's obliviousness about penicillin came to an abrupt halt, in a gruesome fashion.

The Cocoanut Grove, on Piedmont Street in Boston's Back Bay neighborhood, had been a speakeasy through the years of Prohibition, and, after the repeal of the Volstead Act, the city's most popular nightclub. On November 28—the Saturday of the Thanksgiving holiday weekend—more than a thousand partiers crowded into the club, which the Boston Fire Department had authorized for fewer than five hundred. Sometime after 10:00 P.M., one of the artificial palm trees used to decorate the club caught fire, and flames quickly spread to the club's wall and ceiling decorations. In five minutes, the nightclub was an inferno of superheated air and toxic smoke. Panicked people were crushed at exit doors. Four hundred ninety-two people died.

Hundreds more were victims of smoke inhalation and second- and third-degree burns. More than a hundred were brought to Massachusetts General Hospital, whose emergency physicians decided against debriding the burns—picking off particles of clothing or other foreign material; they were convinced that debriding would be more likely to remove what was left of the skin's protection against invading bacteria. Instead, they hoped to fight infection using "chemotherapeutic agents administered internally." Sulfa drugs were available, but only effective against a limited range of infections. It was an emergency, and it demanded an emergency response. Only days after the fire, Merck's Rahway facility packaged a quantity of penicillin—not Anne Miller's powder, but thirty-two liters of highly diluted broth—and transported it, complete with police escort, to Mass General.

It's hard to know whether the miracle medicine performed miraculously. The emergency room physicians at Mass General and the other Boston hospitals used a variety of techniques in treating burn patients, and while dozens survived staph infections that would have been fatal only a few years before, there is no way of knowing for sure whether the broth from Merck was the reason. Victims were treated with protective dressings and given antibacterial treatments—including sulfa drugs—internally. That didn't stop the December 2 issue of the *Boston Globe* from proclaiming penicillin "priceless," and if anyone was disposed to argue the point, no record of it survives. What did survive was the belief that a wonder drug was, at most, only weeks away from widespread use. Before the Cocoanut Grove, the number of human beings that had been treated with penicillin was measured in the dozens. Alfred Newton Richards, chairman of the CMR had, as a matter of policy, restricted the amount of publicly available information on the new drug, offering updates only in vaguely worded press releases. His logic was obvious enough; given that the national priority was manufacturing enough penicillin for battlefield use, raising hopes about its availability to the nonuniformed public would be both cruel and a public relations nightmare. Chester Keefer of Boston University, the chairman of the Committee on Chemotherapeutics and Other Agents, was actively hostile to journalists, whom he believed would inevitably create expectations that could not be met. For months, all communiqués about the penicillin project had been run exclusively through the CMR press office. A single night in Boston and the secret was out. *Time* magazine's February 8, 1943, issue said it all: "The wonder drug of 1943 may prove to be penicillin, obscured since its discovery in Britain in 1929, only now getting its thorough sickroom trial."

Keefer and Richards did their best to keep the lid on, but a sampling of newspaper headlines that followed Cocoanut Grove will give the flavor:

7 HOURS TO LIVE—SCARCEST DRUG RUSHED TO BABY

GIRL, 20, DEAD AFTER REFUSAL OF PENICILLIN

PENICILLIN . . . DEW OF MERCY*

* This was an actual ad in *Time* magazine from the York Refrigeration and Air Conditioning Company touting the importance of refrigeration in preserving the drug for use.

A woman in Oklahoma City wrote to President Roosevelt because "I do not know who else has the authority to help me, or if you can possibly tell me where my son can get the medicine penicillin. . . ." Another mother wrote the president a letter in which she allowed that "I know you are busy with the war, but . . . I am in great need of your help. My husband is in great need of the new drug penicillin. . . ." In the face of such demand, the *New York Herald-Tribune* actually provided hopelessly optimistic instructions for making penicillin in a home kitchen.

The result was that demand and supply were seriously out of balance. During the first five months of 1943, American production of penicillin was only 400 million units; and, since more than 2 million units were needed to treat simple staph infections, the total supply was sufficient to treat fewer than a hundred patients. For skin infections like the acute *streptococcal* skin disease known as erysipelas, a single treatment could require 9 million units or even more: 200,000 to 400,000 units three times daily for ten days.

Penicillin manufacturing had to be industrialized.

In June 1943, Richards and Keefer attended a meeting in Washington hosted by Elihu Root of the National Academy of Sciences. In attendance were Robert Coghill from the Northern Lab, and members of the War Production Board, which had unprecedented authority over all allocations of resources—private and public—for the duration of the conflict. Three months later, a slightly larger version of the same group met again, this time as the WPB's Penicillin Producers Industry Advisory Committee. Two key items were on the agenda. The first was the appointment of a "penicillin czar," formally the coordinator of the penicillin program: Albert Elder, a chemical engineer from the U.S. Patent Office's Chemical Division. The second was to recruit a sufficient number of qualified American corporations to ramp up penicillin production.

The new agenda demanded a wider ambit. At the meetings called by Richards in October and December 1941, only four pharmaceutical companies had been represented: Merck, Pfizer, Squibb, and Lederle . . . and only the first three agreed to commit any resources to the project. By 1943, matters had changed dramatically. The penicillin project was now a national priority, and virtually every company that had anything to do with medicinal compounds, or even fermentation, was invited to apply for consideration, most of them given only the vaguest knowledge of the nature of the project.

From the 175 companies that applied, Richards, Elder, Keefer, and Coghill selected seventeen. Some were obvious, like the first three: Merck, Squibb, and Pfizer. The others included drug companies like Lederle, Eli Lilly, Sharp & Dohme, Abbott Laboratories, Parke-Davis, Winthrop, Upjohn, Cutter Laboratories, Roche Nutley (the American subsidiary of the Swiss drug company Hoffmann-La Roche, located in Nutley, New Jersey), and Bristol-Myers's Cheplin Laboratories; but also companies with experience in fermentation for other uses, such as Allied Molasses, Schenley Distillers, the Heyden Chemical Corporation (like Merck, a once-upon-a-time German-owned firm, seized by the Office of the Alien Property Custodian during the First World War), and the Commercial Solvents Corporation.* Each was promised free access to all publicly available information about penicillin fermentation, *plus* patentable ownership of any techniques they developed while part of the program.

By any standard, this was a dramatic change. As later recalled by Sir Robert Robinson:

> Richards . . . *recognized the simple truth that the commercial interests would not develop penicillin unless they were guaranteed some enjoyment of the fruits of their labors and investment. Somehow the companies that participated in the development of penicillin had to be permitted exclusive rights to their discoveries. Once the research and development of penicillin were completed, no one would be allowed to jump on the penicillin bandwagon for a free ride. The CMR thus*

---

* Under the heading, "strange connections": Commercial Solvents's business was the manufacture of acetone using a process developed by, and licensed from, Chaim Weizmann, the first president of Israel. Notable by their absence were a number of companies approached by Florey during his U.S. tour, such as Smith-Kline, Connaught Laboratories, Lambert Pharmacal, and Mulford.

found itself in the awkward position of needing to devise a system by which private companies would gain patent rights to processes and products developed, at least in part, with public money.

By the end of 1943, the CMR was spending public money like water. The scope of the effort was really stunning in retrospect: The office had recruited thirty-six universities and hospitals, twenty-two separate companies, and four federal, state, local, and national organizations. Some were multimillion-dollar corporations, like Squibb, but not all; the first company to make a real contribution to the effort was the relatively tiny Chester County Mushroom Laboratories of West Chester, Pennsylvania, which was processing forty-two thousand surface cultures a day. In 1943 alone, the CMR approved fifty-four contracts totaling more than $2.7 million for research on penicillin, and agreed to pay penicillin producers $200 for each million units (Chester Keefer was allocated $1.9 million to buy the stuff needed for his clinical trials). In addition, the War Production Board approved sixteen new penicillin-manufacturing plants, on which the pharmaceutical companies spent nearly $23 million. They also sweetened the deal by offering, as a wartime priority, so-called certificates of necessity, along with tax breaks that allowed companies like Merck and Pfizer to depreciate their investments in only five years. The WPB also spent nearly $8 million of federal money on six penicillin-manufacturing plants, all of which were sold to private companies after the war ended, as designated "scrambled facilities," a term of art for assets in which the private and public investments were almost impossible to disentangle— a metaphor for the entire penicillin project.*

---

* After the war, the newly enriched pharmaceutical companies, having gotten a taste for public-private partnership, spent an additional $11.6 million buying or leasing war surplus plants in order to convert them to antibiotics manufacture. Eli Lilly bought an airplane propeller plant. Pfizer a submarine repair base in Groton, Connecticut, and an ammunition loading plant in Terre Haute—648 buildings on more than 6,000 acres—and converted it to the manufacture and packaging of antibiotics. Merck bought the Cherokee Ordnance Works at Danville, Pennsylvania.

It was, to free-market purists, either the greatest of heresies or—more likely—the source of much cognitive dissonance. At the end of the 1920s, pharmaceutical development and manufacturing was the sixteenth most profitable industry in America. By 1944, it was, by far, the most profitable. It would remain so for nearly twenty years.

Moreover, the industry, which had been made up of hundreds of firms, none possessing more than 3 percent of the national market, had consolidated into twenty or so companies that held, in the aggregate, 80 percent of the market for *all* drugs, and that market had grown tenfold. What separated the twenty winners from everyone else was possession of a penicillin contract: Each firm that got an OSRD manufacturing contract quickly outstripped its peers; one contract, in economic terms, was the equivalent of finding three hundred additional researchers or $10 million in profit (the entire profit for a company the size of Squibb in a good year prior to the project). It is almost impossible to overstate the importance of this. A mediocre company—measured by profitability or growth—without an OSRD contract was transformed into one of the most profitable simply by winning the CMR sweepstakes. It was equivalent to giving the winners a two-decade head start on the rest of an entire industry. The only comparable events in American economic history were the deals that built the transcontinental railroad and allocated the radio broadcast spectrum.

The penicillin project would prove a game changer for companies like Merck and especially Pfizer, which staked its future on penicillin. When Jasper Kane, who had represented Pfizer at the October 1941 meeting of the OSRD, brought his plan for fermenting penicillin in the same sort of deep tanks the company used for producing citric acid to his boss, Pfizer president John L. Smith, he was asked, "Is it worth it?" Smith explained, "The mould is as temperamental as an opera singer, the yields are low, the isolation is difficult, the extraction is murder, the purification invites disaster and the assay is unsatisfactory. Think of the risks and then think of the expensive investment in big tanks—think of what it means if you lose a 2,000-gallon tank against what you lose if a flask goes bad."

Kane's reaction is unrecorded, but it must have been persuasive.

Smith had a friend whose daughter had been cured of erysipelas by a series of penicillin injections, and Smith hadn't forgotten. In early 1943, he charged his chief engineer, John McKeen, with building a pilot penicillin plant in an old ice factory and, once the plant had been fitted with fourteen 7,500-gallon fermenting tanks, "followed every tank, every potency and everything on a day to day basis."

He had to. Separating high-grade penicillin from all the other components in the fermentation soup was hard enough in the Northern Lab's washing machine–sized tanks. To scale up production by several orders of magnitude, McKeen hired a process-engineering firm that was experienced with another separation challenge: turning crude oil into petroleum, kerosene, and aviation fuel. Within months, Pfizer's subcontractor, the chemical engineering company E. B. Badger & Sons ("Process Engineers and Constructors for the Petroleum, Chemical, and Petro-Chemical Industries"; their lead engineer, Margaret Hutchinson Rousseau, had been the first woman to receive a doctorate in chemical engineering from MIT) had solved the problem of distributing sterile air throughout the fermenting liquid evenly, by introducing pressurized sterile air at the tank's bottom, while agitators mixed it evenly into the broth. By June 1944, penicillin production exceeded 100 billion units a month, and the dramatic increase in supply had a predictable impact on price. The United States government had, the year before, guaranteed the members of the penicillin consortium that the price for the drug would be set at $200 per million units. When the price supports were removed, the market was able to find a price where supply met demand: $20 for a million units of penicillin . . . on its way to $6.*

For a time, Britain was able to keep pace with American developments, though the spigot through which research money flowed

---

* At around the same time, another member of the penicillin consortium, Eli Lilly, converted its own penicillin production "line" from 175,000 two-quart milk bottles to a 3,000-gallon fermentation tank. The first tankful of penicillin was achieved on January 1, 1944.

remained astonishingly narrow. Research funds continued to depend on patrons with titles. In March 1943, Florey persuaded Lord Nuffield (the former William Morris) to donate £35,000, but only over seven years. Government funds matched this, once Prime Minister Churchill concluded that the Allies might be short of the drug on D-day (or worse, that the entire supply would be reserved for American and Canadian troops), but it is scarcely surprising that British production was severely challenged by the need to keep pace with the Americans.

Two consortia had been selected by the British government to lead the way in penicillin manufacture. The first was the giant Imperial Chemical Industries, which had been able to produce only a few dozen doses of penicillin a week as late as 1942, but had agreed to invest £300,000 in a new state-of-the-art facility (apparently after touring Pennsylvania's Chester County Mushroom Laboratories). The other was the Therapeutic Research Corporation, which had been formed in 1941 as a joint venture composed of the Boots Pure Drug Company,* British Drug Houses, Glaxo Laboratories, British Drug Company, May & Baker, and the Wellcome Foundation.

Because of the support of the prime minister, and heroic efforts on the part of British pharmaceutical firms, British production of penicillin managed to match American output through 1943, though, as noted, the total quantity produced by each country was barely enough for clinical trials.

Britain's contribution to basic research, however, remained critical—Dorothy Crowfoot Hodgkin's in particular. Somehow, Hodgkin had gotten access to the most powerful computing machinery then available, the punch-card calculators that were used by the Royal Navy to assemble the most efficient convoys for transatlantic duty, and by the RAF for bombing tables. The computational work wasn't easy or cheap—the Medical Research Council questioned her bill for

---

* As early as 1930, Fleming had given a sample of his broth to C. E. Coulthard, Boots head of bacteriology, though they, like everyone else, had failed to produce anything useful from it.

computing, convinced it was a mistake; she assured them it was not. But by May 1943, she reported, "Our analysis reached a stage at which we felt reasonably confident that we had found the atomic positions within the crystal structure of . . . penicillin." The dispassionate prose of scientific writing underplays both the scale and importance of Hodgkin's achievement. Without a clear picture of the molecule, attempts to synthesize it—that is, build it from a simpler set of component parts rather than cultivate it in fermentation tanks—were doomed. Improving the antibacterial properties of the compound, however it was produced, was hostage to a clear picture of its structure in three dimensions; if penicillin fought Gram-positive bacteria by disrupting cell walls (it did), some of its molecular components had to be able to latch onto the surface of a pathogen, which required an understanding of where exactly those components were located.

Even more impressive, Hodgkin had used a remarkable technique for elucidating the structure of penicillin; not by taking a picture of it, but by calculating its atomic positions from empirical knowledge of its activity viewed through the lens of very sophisticated mathematics . . . the biological equivalent of finding an otherwise invisible planet by measuring the effects of its mass on other, visible, objects.

However, even with the results of the Fourier analysis of the X-ray crystallography, the molecular structure of penicillin remained controversial.* Robert Robinson, Hodgkin's onetime tutor and now at the Dyson Perrins lab, proposed a structure based on the chemical compound oxazolone. Hodgkin, knowing how difficult it had been for Chain and Abraham to work with the compound, thought oxazolone too stable, and proposed that it was more something else, one that Abraham and Chain had already suspected, and that they therefore "immediately accepted."

* By July 1943, back in the United States, Squibb had managed to crystallize penicillin itself, as a sodium salt, rather than one of its degradation products, as had E. P. Abraham at the Dunn School. Abraham's sodium penicillin was the more complex of the two, but had been isolated first, and so was named penicillin F. The Squibb salt, penicillin G, was grown from the mold found on Peoria's cantaloupe, and was already becoming the dominant variant of penicillin.

The "something else" was a beta-lactam ring.

A beta-lactam ring is a fairly simple chemical feature: a square formed by three atoms of carbon, one of them connected to a doubly bound atom of oxygen; and one of nitrogen, connected directly to the oxygenated carbon atom. Because two of the carbon atoms and the oxygen are bonded together at one angle, and the third carbon is attached to the square at a different angle, the square it forms is constantly under tension—imagine trying to build a square out of struts that are bending away from one another. This gives the ring both its instability—which had frustrated everyone from Fleming to Chain—but also its effectiveness. As early as 1940, researchers from Florey's team at the Dunn had been observing penicillin's activity against pathogenic bacteria, and reported that it didn't kill them or dissolve them immediately; rather that the microbes exposed to penicillin went through the same first stage of mitotic division as other bacteria—elongation—but instead of dividing, they just kept elongating (sometimes ten or twenty times their normal length) until they exploded.

Now, they knew why. All that strain from the different attachment angles made the beta-lactam ring vulnerable to breakage, and the bond that typically broke first—between the oxygenated carbon and the nitrogen atom—ended up adhering the oxygenated carbon to the enzyme needed to create the substance used to make the cell walls of Gram-positive bacteria, the ones that are unprotected by the lipopolysaccharide outer membrane. When the enzyme was locked up by the now-open beta-lactam ring, it couldn't produce a sufficient quantity of the key component of cell walls, so when the cells divided, the new walls were, metaphorically, missing a lot of bricks, and even more mortar.

Unsurprisingly, such walls eventually collapse. Though the debate over penicillin's structure would continue until 1945, when both Hodgkin and the American chemist Robert Burns Woodward were able to produce incontrovertible X-ray crystallographic proof of the beta-lactam ring, the puzzle had, for all intents and purposes, been solved. As a commemorative gift, Hodgkin later presented Chain with a model decorated with pushpins stuck in place to represent the molecule's structure.

Hodgkin's discoveries were exemplary science, just as she was an exemplary scientist. In addition to the Nobel Prize, she was awarded the Order of Merit, the Copley Medal, the annual medal of the Royal Society, and has appeared not once, but twice, on British stamps.* She derived the structure of some of the most medically significant compounds in the history of medicine, including insulin, vitamin B_{12}, and not merely penicillin, but the entire family of antibiotics. At her memorial service in 1994, the molecular biologist Max Perutz, a colleague at both Oxford and Cambridge, said, "She radiated love: for chemistry, her family, her friends, her students, her crystals, and her college. . . . There was magic about her person. She had no enemies, not even amongst those whose scientific theories she demolished or whose political views she opposed."

Applying Fourier analysis to X-ray crystallography—to help to synthesize penicillin—was to prove both elusive and an expensive lure for the American and British penicillin efforts. For academics like Dorothy Hodgkin and corporations like ICI and Pfizer, the manufacture of penicillin by fermenting mold exudate seemed slightly disreputable; a temporary stopgap at best, something like treating malaria by boiling cinchona bark to get quinine rather than taking Atabrine tablets. It wasn't merely that growing medicines, at that moment in history, seemed a bit medieval. It was also far more difficult to standardize dosages, or even, as researchers at the Northern Lab had learned, to find a reliable "pure" strain of *Penicillium*. Out of a combination of practicality and pride, Merck alone invested nearly $800,000 in synthesis experiments through 1944, and even promised

---

* In the great though occasionally embarrassing tradition of English scientists with leftist politics, she was also a recipient of the Lenin Peace Prize and the Lomonosov Gold Medal, awarded by what was then the Soviet Academy of Sciences. One consequence is that she was frequently barred from traveling to the United States. Just to keep everyone on their toes, however, her most famous student, Margaret Thatcher, née Roberts, was a lifelong friend who kept Hodgkin's portrait in her Downing Street office.

Vannevar Bush at the OSRD—which had provided American universities grants amounting to an additional $350,000 to investigate penicillin synthesis—a bottle of synthetic penicillin by the beginning of 1945. Chemical synthesis was clearly seen as a superior, *modern*, way to make medicine.

There was another, more urgent, reason to master the technique. The Allies were in a shooting war with the world's best chemical synthesists: the Germans, who had first synthesized mepacrine/Atabrine as an antimalarial in 1931. And yet, despite the undeniable fact that the academic and industrial resources of Germany, at least as regarded chemical innovation, were superior to those in the United States or the United Kingdom before the war, they never developed a wartime antibiotics program. The question is, why not?

At first glance, the answer might appear to be the enormous resources commanded by Vannevar Bush, Alfred Newton Richards, and the OSRD. Or, for that matter, the industrial strength that gave the United States, by the end of the Second World War, nearly half of the entire world's gross domestic product. It was probably inevitable that American infrastructure would eventually dominate the new business of drug production, as indeed it did in every other measurable economic activity.

Even so, that explains American *postwar* dominance a whole lot better than the achievements that occurred *during* the war. Germany possessed an enormous head start in the key industries of drug development and production, was preeminent not only in every aspect of chemical manufacturing, but also benefited from hundreds of alliances between commercial enterprises like the I. G. Farben cartel and what were, at the outbreak of the war, still the world's most prestigious universities. Moreover, once the Oxford group started publishing in 1940, academic papers about penicillin started appearing practically every week in both English and German, which means that however much the OSRD and Britain's Medical Research Council tried to keep the details secret, by 1942 the *Penicillium* cat was out of the bag. And yet, even by 1945, Germany was able to produce only about 30 grams of penicillin a month, no more than the quantity required to treat four dozen or so patients. Germany, where both Salvarsan and Prontosil were introduced, had become a dead end in the search for more powerful anti-infective treatments.

The reason certainly wasn't because the Nazi state lacked an urgent need for treating battlefield injuries. In May 1943, when the OSRD and the War Production Board approved an additional sixteen new plants for producing the penicillin needed for D-day, thirty thousand soldiers of the Wehrmacht died on the eastern front alone, a huge number from septic wounds.

So far as can be gleaned from the historical record, the answer is not primarily, "They were manufacturing Zyklon B for the gas chambers instead." Mass killing on an industrial scale was, indeed, a national priority for the Nazi state; but, for Germany's great chemical companies, an even higher priority was oil. The Saar region had enough coal to fuel the Industrial Revolution, and more than enough to run German factories. Oil, though, particularly petroleum, was a different matter. Outside of Romania, there wasn't a decent-sized oil field anywhere from the Atlantic to the Urals. Which was why, even before the Nazis took power in 1933, I. G. Farben was investing enormous resources in the manufacture of synthetic fuels: $100 million and $125 million in current dollars between 1925 and 1932, or at least $1.7 billion today.

It wasn't, by traditional standards, a profitable investment. The Leuna brand of synthetic gasoline—the name came from the facility where it was produced, in the Saxon city of Leuna, near Leipzig—was an attempt to gasify Germany's still-abundant reserves of coal, using the chemical process known as hydrogenation. Even with substantial subsidies from the Weimar government, though, it hemorrhaged red ink from 1930 forward, and continued to do so after the Nazis took power in 1933. Carl Bosch, the Nobel Prize–winning head of BASF and, since the 1925 merger, a director of I. G. Farben, instructed his staff to provide documentation for even larger state subsidies, projecting that the German state would consume 50 percent more fuel oil and petroleum by 1937. Bosch's protégé and successor, Carl Krauch—

a Nazi Party member, unlike his anti-Nazi boss—proposed to Hitler's cabinet that domestic production of fuel oil and petroleum could be increased between 25 and 63 percent, from 500,000 tons to nearly 3 million tons annually. If, that is, the national government could close the German market to "foreign influences" and agree to buy the fuel at a substantial premium over the world market price.

What this meant, in effect, was that the Nazi state would be subsidizing the production of oil to the extent that nearly 40 percent of *all* industrial investment in Germany before 1939 (except for coal and electricity) went to either synthetic oil or—the other requirement of a mechanized army—synthetic rubber.

It was a success, if that's the correct word, for I. G. Farben. Revenues soared, at least partly due to a gruesome policy for controlling labor costs: Dozens of Farben's directors would serve time as war criminals for employing slave labor in its synthetic oil and rubber plants. Scholven AG, a joint venture with a number of German mining companies, produced 125,000 tons of synthetic fuel in 1936; by August 1939, just before the start of the Second World War, I. G. Farben had twelve hydrogenation plants making gasoline and other refined oil products, for which they earned between $65 million and $140 million annually—up to $1.5 billion in current dollars. Another $50 million was earned annually producing synthetic rubber.

This also guaranteed, though, that tens of millions of dollars *weren't* being spent subsidizing research into any drugs, much less penicillin. Perversely enough, pharmaceutical companies in the United States and the United Kingdom, fearful of investing in factories that could be made obsolete so quickly, diverted millions of dollars from a technology that actually worked—fermentation—in an attempt to surpass Germany's perceived mastery in chemical synthesis. And they did so while the Germans were embarked on an entirely different project. The country that had the world's best chemists in the 1930s directed them to spend the decade—and a ridiculously large percentage of the nation's investment capital—not in pharmaceutical innovation, but in supplying fuel to the Wehrmacht. The big difference was that the United States was wealthy enough to afford to make expensive mistakes. Germany wasn't.

In 1943, British production of penicillin had been approximately equal to that of the United States. In 1944, it was barely one-fortieth as large.

Howard Florey, more responsible than anyone for that extraordinary achievement, spent most of 1943 and 1944 in the field, investigating how best to use penicillin for treating battlefield injuries. On a trip to his native Australia, he gave forty-two lectures on the proper use of the drug, and would eventually train five hundred clinicians and more than two hundred pathologists on penicillin therapy.* He also demonstrated the effectiveness of penicillin in treating gonorrhea, which was believed to be at least as dangerous to Allied troops as German artillery. One of the largest grants from the CMR during the months leading up to D-day had been to research the best ways to use penicillin to help treat gonorrhea—one American administrator noted that "the goal [is] to make penicillin so cheaply that it costs less to cure [VD] than to get it . . ."—thus keeping tens of thousands of troops at least putatively battle ready (though also creating an ethical dilemma about treating civilians in postwar Europe in preference to STD-infected soldiers).

By December 1945, when Florey, along with Alexander Fleming and Ernst Chain, received the Nobel Prize in Physiology or Medicine, penicillin had already saved tens—perhaps hundreds—of thousands of lives. But it transformed the world in other ways, too. The penicillin project had created an entire industry, and built what would become some of the most profitable companies in history: not merely

---

* Florey's innovation, "instillation therapy," called for closing infected soft tissue wounds as fast as possible, leaving rubber tubes for irrigating the site with penicillin. Not until the end of 1944 was enough penicillin available to replace instillation with intramuscular injection.

the American participants in the penicillin project, but also British firms like Glaxo, France's Rhône-Poulenc, and even Swiss companies like CIBA-Geigy and Sandoz. At the Nobel Banquet, Professor A. H. T. Theorell of the Nobel Institute of Medicine toasted the laureates thus:

> *To you, Ernst Chain, Howard Florey, and Alexander Flem-*
> *ing, I will relate one of Grimm's fairy-tales, that I heard as a*
> *child. A poor student heard under an oak a wailing voice that*
> *begged to be set free. He began to dig at the root, and found*
> *there a corked bottle with a little frog in it. It was this frog*
> *that wanted so badly to be set at liberty. The student pulled*
> *the cork, and out came a mighty spirit, who by way of thanks*
> *for the help gave him a wonderful plaster [i.e., bandage]. With*
> *the one side, one could heal all sores; with the other one could*
> *turn iron into silver. . . .*

Florey, Chain, and Fleming, along with a long list of colleagues, hadn't quite healed all sores with their discovery. Penicillin was widely, but not universally, effective; it had no curative powers against infectious diseases caused either by viruses or by Gram-negative bacteria. But for institutions like Merck, Pfizer, Squibb, and all the others, it had indeed turned iron into huge quantities of silver. The first antibiotic, and its successors, would do so for many decades to come.

SIX

"Man of the Soil"

In November 1915, Goodhue County, Minnesota, opened the Mineral Springs Sanatorium ("for consumptives") at Cannon Falls, a small town about thirty-five miles south of Minneapolis. Mineral Springs had only thirty-four beds when it opened, a small hospital, even by the standards of the rural Midwest. Forty years later, the number of beds had increased, but the disease that filled them, year in and year out, was always the same: the "white plague"—pulmonary tuberculosis.

One of those beds, in October 1945, was occupied by a twenty-one-year-old woman believed to be days away from death. It's unknown how she contracted pulmonary tuberculosis, but the overwhelming likelihood is that she breathed in aerosol droplets that had previously occupied the respiratory tract of another infected person. Perhaps that person sneezed, or coughed, or just exhaled. It didn't matter; while a sneeze can transfer a million cells of the bacterium known as *Mycobacterium tuberculosis*, ten is all it takes to start a colony. The young woman, Patricia Thomas, had definitely been colonized. She had endured months of high fever, loss of appetite, weight loss, muscle atrophy, and the disease's chronic, blood-tinged cough.

In the year of her hospitalization, seventy-five thousand Americans would be killed by the deadliest infectious disease in history. By then, tuberculosis had killed one-seventh of all the humans who had ever lived: more than fifteen billion people.

That much devastation leaves a trail. Skeletons exhumed from Egyptian gravesites more than four thousand years old have the characteristic deformities of Pott's disease, or spinal tuberculosis, as do skeletons from Neolithic sites in northern Europe and the Mideast. Almost from the moment human civilizations began to record the stories of their lives, tuberculosis was a featured player, generally appearing in the last chapter. Assyrian clay tablets describe victims coughing blood before they died. Hippocrates treated patients wasting away with chest pain and drowning in bloody sputum. Ancient Chinese physicians called the disease *xulao bing*; Europeans, consumption. Recommended treatments had included insect blood, breast milk, bloodletting, high-altitude living, sea travel, drinking wine, avoiding wine, and—most famously—the "king's touch," which was the belief that a divinely sanctioned monarch, such as the king of England or France, could, by a laying on of hands, cure what they knew as scrofula.*

However, the recorded history of tuberculosis is only the last part of its story. For a long time, the accepted wisdom was that tuberculosis emerged around the same time that humans discovered agriculture and started forming settled communities, during the so-called Neolithic Demographic Transition, or NDT, which began some ten to twelve thousand years ago. The reason that medical historians assumed this for so long has to do with the complicated nature of tuberculosis itself. Most of the really scary infectious diseases are what demographers call "crowd diseases." Crowd diseases tend to feature high mortality—up to 50 percent when untreated—and a very rapid means of transmission. From an evolutionary perspective, the two go hand in hand; pathogens that kill their hosts need to spread rapidly, or die out. Most crowd diseases are known to have emerged around the time of the NDT, when *Homo sapiens* not only experienced a rapid population explosion but an even faster increase in population density. Villages and cities, where people and domestic animals lived in far closer quarters than the nomadic hunter-gatherer societies they succeeded, created a target-rich environment for crowd disease

---

* The proper name for scrofula, a disease that manifests with swollen and eventually ulcerous growths on the lymph nodes of the neck, is tuberculous cervical lymphadenitis, a form of tuberculosis that presents outside the lungs.

pathogens. For that reason, a large number of crowd diseases started as zoonoses: animal diseases that were able to jump to human hosts.

But while tuberculosis is both highly dangerous and very easily spread, it also has the characteristic features of chronic, noncrowd diseases, the ones caused by pathogens that thrived in the low-density societies that preceded the NDT. Most such pathogens were not originally zoonotic, and are typically able to lie dormant within a host for years, only to be reactivated when opportunity presents. *Treponema pallidum*, the bacterium that causes syphilis, is a famous example, as is *Borrelia burgdorferi*, the bacterium behind Lyme disease. Another is *Mycobacterium leprae*, which is the causative agent for leprosy; and predictably enough, so is its cousin, *M. tuberculosis*. One implication of this fact is that tuberculosis is not, as once was thought, a disease that jumped from domestic animals to humans. Since it's older than domestication, it's far likelier that humans originally gave it—the version that is now known as *M. bovis*—to farm animals, rather than the other way around. It certainly appeared in human populations long before our ancestors first experimented with animal husbandry. Decades after Patricia Thomas encountered *M. tuberculosis*, genetic analyses of hundreds of different strains of the bacterium decided the issue: The pathogen had originated in Africa, somewhere between forty and seventy thousand years ago.

In the same way that *H. sapiens* dispersed, evolved, and expanded, so too did *M. tuberculosis*: from Mesopotamia eight to ten thousand years ago, to China and the Indus valley a few thousand years later, eventually to Europe and the New World. It didn't, of course, stop there. *M. tuberculosis* experienced another evolutionary burst only a few centuries ago. The most successful modern lineage—the so-called Beijing lineage—experienced a population increase over the last two centuries almost exactly paralleling the sextupling of human population during the same span.

A population of pathogenic bacteria that seemed always to be growing larger was bad news for its human hosts. But it wasn't quite bad enough. As the pathogen was increasing its numbers, it was also evolving new and deadly virulence factors, which is one reason that

Patricia Thomas was dying in her hospital bed. Though M. *tuberculosis* has been, by the standards of bacteria, an extremely slow-evolving and stable organism—the ubiquitous bacterium known as *E. coli* divides and replicates about once every twenty minutes, while *M. tuberculosis* does so only every fifteen to twenty *hours*, which means that spreading an evolutionary adaptation through a population takes fifty times as long. The small portions of its genome that do *mutate* almost always increase its virulence.

Even today, tuberculosis virulence isn't fully understood. However, it is known that unlike most pathogens, including the strep bacterium that nearly killed Anne Miller,* M. *tuberculosis* doesn't produce toxins. Instead, the microbe is ridiculously efficient at hijacking the host's own defenses and transforming them into deadly attackers. Any time the host's immune alarm bells go off, it summons macrophages, the oversized white blood cells, to the site of infection. The macrophages, whose job it is to engulf and digest foreign objects, form cavities: vacuoles known as phagosomes that surround the invading pathogens. Once surrounded, the macrophage then connects the phagosome to the lysosome, a chemical wood chipper that uses more than fifty different enzymes, toxic peptides, and reactive oxygen and nitrogen compounds that can, in theory, turn any organic molecule into mush.

When they attempt this with *M. tuberculosis*, however, things don't work out as planned. The bacterium secretes a protein that modifies the phagosome membrane so it can't fuse with the lysosome. Thus protected, it is able to transform the macrophage from an execution chamber to a comfortable home—one with a well-stocked larder, since another of the pathogen's talents is the ability to shift from dining on mostly carbohydrates (which is what it eats when grown in a Petri dish) to consuming fatty acids, particularly the cholesterol that is a common component of human cell membranes.

It's the replication that matters. Within three to eight weeks after

---

* *Streptococcal* toxins are a terrifying bunch. One of them, known as pyrogenic exotoxin C, causes the distinctive rash of scarlet fever; another, streptolysin, does to cell membranes what boiling water does to a sugar cube.

breathing an aerosol containing a few hundred *M. tuberculosis* bacteria, the host's lymphatic system carries them to the alveoli of the lungs: the tiny air sacs where carbon dioxide is exchanged for oxygen. As *M. tuberculosis* forms its colonies inside macrophages, they create lesions: calcified areas of the lung and lymph node. Some burst; others form a granuloma—a picket fence of macrophages—around the colony. Within three months, the interior of the granuloma necrotizes, that is, undergoes cellular death. Some of the deaths occur within the lungs, leading to the painful inflammation known as pleurisy, which can last for months. Other infested areas, known as tubercles, break off from the lungs and travel via the bloodstream to other parts of the body, becoming the frequently fatal form of the disease that physicians know as extrapulmonary tuberculosis—Patricia Thomas's diagnosis. When it settles in the skeletal system, it can cause excruciatingly painful lesions in bones and joints. When it lands in the central nervous system, as tubercular meningitis, it causes the swelling known as hydrocephalus; on the skin, where it's known as lupus vulgaris, it leaves tender and disfiguring nodules.

And, even when the body's determined immune system destroys most of the granulomas, they leave behind huge amounts of scar tissue, which weaken the host's ability to breathe. Bronchial passages are permanently blocked. Frequently, the cells needed for oxygen uptake are so damaged that victims suffocate. Sometimes the deadliest attacks of all are friendly fire. The immune system's inflammatory response, which evolved to clear out damaged cells and allow rebuilding to follow, can overshoot the mark, especially when confronted with an especially robust (or wily) invader. When it does, histamines and the other compounds that increase blood flow and ease the passage of fluid through cell membranes cause enough fever and swelling to kill hosts as well as pathogens.

Most victims, nonetheless, survive a first bout with tuberculosis, generally because the colonies of *M. tuberculosis* lack the time to achieve maximum size before the host's immune system intervenes. The problem, however, is that the disease doesn't disappear. It stays latent, waiting for the opportunity to start the cycle of replication.

One in ten times the full-blown disease will reappear as secondary tuberculosis, either because of reexposure to the pathogen, poor enough nutrition that the host's immune system is damaged, or even hormonal changes. The result is that the same host of symptoms attacks a much-weakened host within a few years of the initial infection. Which is what happened to Patricia Thomas.

With the disease's talent for hijacking the body's own defenses and capacity for remaining lethal years after it had seemingly been defeated, it's no surprise that physicians have been battling the white plague for as long as there have been physicians. They haven't always accurately identified it, or even agreed with one another about any of its characteristics. Hippocrates thought the disease was inherited; Galen that it was contagious. Fifteen centuries later, the debate still raged. Paracelsus, the Swiss physician whose theory of health depended on the proper balance between mercury, sulfur, and salt, couldn't understand why, if the disease was contagious, so many people in European cities who were exposed to it exhibited no symptoms.

Even after Robert Koch identified the guilty bacterium in 1882, the controversy wasn't settled. One reason that the belief in tuberculosis as a hereditary condition proved so durable was that environmental conditions can, indeed, affect the likelihood of activating a latent infection. Pure mountain air didn't cure TB, but it did seem to slow it down.

As a result, sanatoriums, places where patients could cough out their lives in relative comfort, sprang up all over Europe, especially in regions near the mountains or the sea. "Climate cures" built boomtowns throughout the American West, from Albuquerque to Pasadena to Colorado Springs, each of them advertising themselves as the ideal destination for tubercular patients. Edward Trudeau, an American physician who believed tuberculosis to be hereditary, contracted the disease himself in 1873, and built a European-style sanatorium in Saranac Lake, New York, where residents* were confined to bed. Though

---

* One of them was Robert Louis Stevenson, who spent the winter of 1887–88 at Saranac Lake. His timing was not the best; in March 1888, the Great Blizzard dumped more than four feet of snow on the sanatorium.

not bedrooms. Trudeau, convinced of the curative powers of clean mountain air, required his patients to sleep outdoors, even in subzero temperatures. The Saranac Lake model actually became so popular that, in the first decade of the twentieth century, one American in 170 lived in a self-declared sanatorium.

The idea of housing consumptives together would scarcely have made sense absent the conviction that tuberculosis victims weren't themselves infectious. It was a plausible enough notion. Because the pathogen could establish colonies without—at first—causing much in the way of symptoms, hosts could be exposed to the disease without appearing to contract it. Since living in places where the air was clean (and, more important, the sanitation well managed) seemed to relieve symptoms and even slow the progress of the disease, many physicians thought tuberculosis couldn't be infectious (or, at least, not mostly so). At a time when the germ theory itself was still very novel, nineteenth-century European and American societies were largely ignorant of the dangers of spreading the disease. One consequence was that it became a trope for the nineteenth century's literary romantics. Pale heroines languishing beautifully and consumptive children who "appear like fairies to gladden the hearts of parents and friends for a short season" became notorious clichés of Victorian literature. Nor was tuberculosis of interest only to romantics; the residents of the Berghof atop Thomas Mann's Magic Mountain form a cross-section of early twentieth-century European intelligentsia, united only by metaphysics and obstructed breathing.

Like the European society they symbolized, though, their ailment was undergoing a gigantic transformation. As the germ theory took hold, tuberculosis was transformed from a romantic ailment to a contagious one. Consumptives were no longer regarded as brave symbols of individual suffering, but as a social danger, the modern version of lepers.* Public health campaigns warned everyone to take care around coughing and sneezing, and especially advocated for isolating tubercular patients. By the end of the 1920s, sanatoriums were already

---

* Despite its reputation, leprosy, or Hansen's disease, is several orders of magnitude less infectious than tuberculosis.

being transformed from comfortable resorts for the wealthy to quasi prisons for the poor.

Segregating the infected from the not yet infected was harsh, but nearly the only useful response to the disease. Koch's tuberculin had been both a failure and a scandal; Colenso Ridgeon's cure in *A Doctor's Dilemma* was a fiction. Shaw's play asked audiences if "thirty men . . . found out how to cure consumption . . . Why do people go on dying of it?" Though a tuberculosis vaccine had been developed by the French physicians Albert Calmette and Camille Guérin as early as 1916, and first administered to humans in 1921, it was at best only partly effective.* One popular surgical procedure, the so-called pneumothorax technique, deliberately collapsed an infected lung to allow the lesions caused by the tubercular granulomas to heal. The only characteristic all of these purported cures shared was an almost complete lack of effectiveness. Patients came to sanatoria like Mineral Springs not to be cured, but to be made as comfortable as possible while their bodies repaired themselves, or—more frequently—while they waited to die.

A month into her stay at the Mineral Springs Sanatorium, on November 15, 1945, Patricia Thomas became the first patient to receive an injection of a new compound called streptomycin. After a series of injections of this new compound over the following months, she went home, completely cured.

Her battle with *Mycobacterium tuberculosis* ended, but the war over the discovery of streptomycin would rage on, never really subsiding. Truces are broken regularly, whenever advocates for the two scientists at the heart of the quarrel—Selman Abraham Waksman and Albert Schatz—square off. Entire books have been devoted to

---

* Though the BCG vaccine—for Bacilli-Calmette-Guérin—is today the most widely prescribed vaccine in the world, it has shown only a 20 percent level of effectiveness against pulmonary tuberculosis; its greatest value seems to be in forestalling tubercular meningitis. This is one reason that, in the twenty-first century, *M. tuberculosis* is still resident in more than two billion people, a hundred million of whom will develop the disease.

arguing one side or the other. One of the most prestigious science magazines in the world, *Nature*, was, for a time, a battlefield, when Schatz's champion, the microbiologist and historian Milton Wainwright, fought on the page with Waksman's defender, William Kingston, a professor of business and history.

Here's what isn't disputed in this war for credit: Albert Schatz graduated from Rutgers University in New Brunswick, New Jersey, in May 1942 with a degree in soil science, and immediately started graduate work under one of the field's leading lights, Selman Waksman. Both professor and student specialized in actinomycetes, a suborder of soil-dwelling bacteria with thousands of known members, even in the 1940s.

Their interest was more than simple intellectual curiosity. Ever since the 1920s, actinomycetes, an unusual group of bacteria that exhibit *Penicillium*-like branching filaments, had been identified as powerful antagonists to other bacteria. This wasn't really unexpected, since a teaspoon of soil can be home to a billion or more bacteria, which meant a huge number of competitors in the never-ending hunt for the raw materials essential to produce the biomass needed by all life: DNA, RNA, amino acids, fats, and the like. In a nutrient-rich environment like soil, packed with organic material and trace minerals, the competition is fierce, and actinomycetes were already known to be using some very powerful weapons indeed, each one an enthusiastic killer of other bacteria.

In 1939, the French-born microbiologist René Dubos, a former student of Waksman's then working at the Rockefeller Institute for Medical Research in New York, had isolated two distinct compounds from another family of soil bacteria, *B. brevis*. Each of the compounds, which Dubos named tyrothricin and gramicidin, were, like penicillin, enthusiastic killers of Gram-positive pathogens. Unfortunately, unlike penicillin, they didn't do so by weakening the distinctive Gram-positive cell wall. Tyrothricin blocked the synthesis of proteins; gramicidin made cell membranes permeable to salts. Since animal cells have membranes and depend on protein synthesis, both activities made the compounds nearly as deadly to hosts as to pathogens.

Dubos's discoveries weren't inconsequential; gramicidin is still prescribed today for infections of the skin and throat. But to treat systemic infections like tuberculosis, a drug must enter the blood-stream. Since neither gramicidin nor tyrothricin could do so safely, their largest contribution to the antibiotic revolution was to hint that soil might be valuable for growing more than just crops. Somewhere in the dirt there had to be something that would kill pathogens while leaving their hosts alive.

The pursuit of a substance with the right balance of destructive-ness and discretion was a full-time job at Waksman's lab at Rutgers. Throughout the 1920s and 1930s, Waksman and his team had been collecting different soils from all over the eastern United States, iso-lating the varieties of actinomycetes they contained, and then grow-ing them in Petri dishes filled with agar. Once they had a colony, the graduate students working in Waksman's labs would expose it to an-other sort of bacteria, and note the results; if the newly introduced bacteria failed to thrive, then that particular colony of actinomycetes had an antibiotic property.

This is the most tedious sort of research, completely lacking in bursts of inventive genius or innovative experimental design. It did not, however, lack for institutional support.

In 1939, Merck & Co. had concluded an agreement with Selman Waksman that provided the lab with an ongoing grant for the study of antibiotics. Merck's support included money—Waksman had initially been engaged to consult on microbial fermentation for $150 a month; later that year, Merck added another $150 for working on "antibacterial chemotherapeutic agents," along with experimental animals, and fund-ing for a fellowship in the Rutgers lab. (The investment in the fellowship wasn't purely altruistic; Waksman's first fellow, Jackson Foster, later be-came the director of Merck's microbiological lab.) Waksman would do the research, and in return for the funding, and for handling "produc-tion, purification . . . and to arrange for clinical trials," Merck would own the patents from any resulting research, and would pay a royalty of 2.5 percent of net sales to the Rutgers Endowment Foundation, a non-profit charity originally established to solicit donations from alumni.

For its first year or two, Merck's investment hadn't paid much in the way of dividends. However, at the beginning of 1940, Waksman reacted to news of the progress of Florey's team by saying, "These Englishmen have discovered [what] a mold can do. I know the actinomycetes can do better" and proved as good as his word. By the spring of 1941, he had presented his patrons at Merck with the first fruits of his actinomycete farm: the antibacterial compounds clavacin, actinomycin,* and streptothricin. The harvest was promising if not spectacular. All three compounds were effective, but toxic; strepothricin, in particular, was frustratingly able to kill a variety of Gram-negative pathogens in mice, but had the unfortunate side effect of destroying kidney function in the four human volunteers on whom it was—prematurely, not to say irresponsibly—tested.

Waksman was undaunted. It was around this time that he coined the word by which this variety of antibacterial drugs would henceforth be known: antibiotic, which he defined as "a chemical substance produced by microorganisms"—i.e., penicillin, but not Salvarsan—"which has the capacity to inhibit the growth of and even destroy bacteria and other organisms."

The problem was finding the antibacterial needle in the enormous actinomycete haystack. Years later, Waksman would tell people, "We isolated one hundred thousand strains of streptomycetes [as actinomycetes were then known]. Ten thousand were active on agar media, one thousand were active in broth culture, one hundred were active in animals, ten had activity against experimental TB, and one turned out to produce streptomycin." Though the numbers are casual approximations, the technique was essentially that: Throw lots of actinomycetes up against the wall, and see which ones stuck.†

---

* While actinomycin's cost-benefit relationship was unappealing for the treatment of tuberculosis and most other bacterial infections, it would be used effectively to treat a rare form of childhood kidney cancer known as Wilms' tumor. In a 1983 reminiscence, Max Tishler of Merck recalled being awestruck by seeing half a dozen children at the Dana-Farber Cancer Institute who had been cured by the "commercially unimportant" drug.

† One observer would note that Waksman's "real discovery was not streptomycin; it was the principle that a patient, systematic search for useful antibiotics will eventually pay off."

Which is how Schatz spent his days, from May 1942, when he joined Waksman's lab, to November, when he was drafted to serve in an Army Air Forces medical detachment in Florida. It's what he did during his off hours in Florida, which he spent finding, and sending, different soil samples back to Waksman's lab in New Brunswick. And it's what occupied him after he was given a medical discharge, in June 1943, and returned to Waksman's lab.

Albert Schatz (1920–2005) and Selman Waksman (1888–1973)

He did so as one of the few buck privates in the United States Army who took a cut in pay returning to civilian life. Private Schatz had been earning fifty dollars a month while in the service, along with free housing, food, and clothing. As a civilian PhD candidate, he performed the (literally) dirty work of analyzing soil samples for even less: forty dollars a month, which was well below the minimum wage for a forty-hour week. And Schatz, like PhD candidates then and now, worked a lot more than forty hours. By his own, understandably aggrieved, account, "During the four month interval between June and October, 1943, I worked day and night, and often slept in the laboratory. I prepared my own media and washed and sterilized the glassware I used." He even cadged his meals from the stuff grown in the university's agricultural college. He didn't do so without complaint. But he did it, convinced that actinomycetes held the key to a yet-to-be-discovered class of pathogen killers.

And those too-toxic-for-humans antibiotics like streptothricin killed Gram-negative bacteria. The dissertation that Schatz was researching on a salary of ten dollars a week was explicit: "Two problems, therefore, appeared to be of sufficient interest to warrant investigation; namely . . . a search for an antibiotic agent possessing . . . activity . . . against Gram-negative eubacteria . . . and a search for a specific antimycobacterial agent." The sulfa drugs and penicillin were both enzyme blockers, the first inhibiting the ones bacteria needed to synthesize an essential B vitamin, the second blocking the enzymes needed to assemble the giant molecule of amino acids and sugars that make up the bacterial walls of the Gram-positive pathogens *streptococci, staphylococci,* and *clostridia.* But Gram-negative bacteria, with their very different cell walls, are some of the most prolific killers in human history, including *Yersinia pestis,* the bacterium that causes bubonic plague, and *Vibrio cholerae,* which causes cholera.*

Schatz's phrase "specific antimycobacterial agent" was a euphemism for "TB killer." It was also a red flag for Waksman, Schatz's boss and thesis advisor, who recognized that his soil science lab wasn't a secure research facility for something as dangerous as the white death bacillus. Even so, he agreed to support Schatz's research, with the proviso that it be conducted in an isolated basement that he turned

---

* Mycobacteria, though they don't stain, are actually Gram-positive. However, their membranes, constructed from another substance known as mycolic acid, are just as resistant to penicillin as Gram-negative bacteria like *Y. pestis.*

over to his graduate student, either out of an exaggerated fear of a tuberculosis outbreak—Schatz's recollection—or a perfectly reasonable exercise of caution, since the lab lacked what were, even in 1943, state-of-the-art defenses against disease outbreaks: no ultraviolet lights that could be used to kill dangerous microbes; no negative-pressure fans and air filters to keep them from escaping.

Whatever the reason, Schatz's "exile" (his word) produced results. On October 19, 1943, after exposing hundreds of actinomycetes from different soils to the most virulent of tuberculosis bacilli, a variant of *M. tuberculosis* designated H-37, he found two that were antagonistic, one taken directly from the soil, the other swabbed from the throat of a chicken. Both were variants of a bacterium known since 1914 as *Actinomyces griseus*. Schatz renamed it *Streptomyces griseus*. No one knows who named the substance they produced; the first mention of streptomycin in print was in a letter from Selman Waksman to Randolph Major at Merck.

Here was the first step in finding a practical antibiotic. The next one, which Florey's team at Oxford had faced three years before, was running a series of trials to find if it worked not just in vitro, but in vivo—in living, breathing animals. As with the first penicillin extracts, this required streptomycin to be produced in quantity. A sufficient amount for one-off experiments could be distilled in Waksman's New Jersey lab. Or, it could as long as Schatz was willing to be Waksman's Norman Heatley. He set up a similarly makeshift system in his basement lab; to prevent the liquid from boiling away while he slept, the night porter at Waksman's lab awakened him whenever the liquid fell below a red line Schatz had drawn on the distillation flasks.

However, testing Schatz's new compound for effectiveness in vivo needed a more sophisticated facility than the young biologist's basement lab could provide. And in 1945, there was no research hospital in America more sophisticated than the Mayo Clinic. Founded in 1846 by an English émigré doctor named William Worrall Mayo as an outpost for what was literally frontier medicine, the clinic moved to Rochester, Minnesota, in 1864, where Mayo joined the U.S. Army Medical Corps as a member of the state's enrollment board, which examined recruits for the Union army.

From the time Mayo's sons William J. and Charles H. joined him, the clinic was leading the transition from medicine as art to medicine as science. And, as the scientific advances of the nineteenth century were, by the beginning of the twentieth, almost entirely the result of collaboration between specialists, so too was medicine at the Mayo Clinic. By 1889, when the Mayos joined with the Sisters of St. Francis to build Saint Mary's Hospital, William J. Mayo was writing, "It has become necessary to develop medicine as a cooperative science: the clinician, the specialist, and the laboratory workers uniting for the good of the patient. . . . Individualism in medicine can no longer exist." When William Mayo first expressed this sentiment about "cooperative science," he was really writing about a perceived deficiency in clinical practice, not medical research; he wanted to apply the best features of the latter to the former, which, in part, explains the decision to reconfigure the clinic as a not-for-profit charity in 1919. Now, staff would be salaried, not contract, physicians. This, the Mayos believed, would encourage collaboration between specialized researchers and clinicians, who would no longer run the risk of financial penalties for enlisting the best research in their practices.

This forward-thinking philosophy would transform the Mayo Clinic into a world-famous research laboratory. The transformation was already complete when the top veterinary pathologist in the country, William Feldman, joined its Institute of Experimental Medicine in 1927. His research partner, Corwin Hinshaw, had an even more unusual background: Before receiving his medical degree in 1933 from the University of Pennsylvania, he had already done graduate work in zoology, bacteriology, and parasitology. What the two had in common was an interest in lung disease, particularly bovine, avian, and especially human tuberculosis.

By the middle of 1943, Feldman and Hinshaw had performed experiments testing a variety of compounds, including several of the sulfa drugs, and a related class of drug known as the sulfones on tuberculosis in guinea pigs. Sulfones had shown some promise in treating another mycobacterium-caused disease—leprosy—and the Mayo team reasoned, *eo ipso*, that they might be effective against tuberculosis. They were to be disappointed; the curative powers of antileprosy medications—Promin (from Abbott Laboratories) and Promizole (from Parke-Davis)—against *M. tuberculosis* weren't completely absent, but nearly so.

Far more encouraging was Selman Waksman's work with actinomycetes. After reading his papers on streptothricin, Feldman visited Rutgers in November 1943, and encouraged Waksman to keep the Mayo team in mind for anything that showed streptothricin's antibacterial effectiveness without its very high cost in toxic side effects.

In February 1944, Feldman and Hinshaw received an advance copy of Schatz's first streptomycin paper. Though the paper listed twenty-two different bacteria that were either killed or inhibited in vitro by the newly discovered compound, they saw only one: *M. tuberculosis*. The following month, Waksman wrote a letter to Feldman asking whether he and Hinshaw would be able to perform clinical tests in vivo on the drug.

It took Schatz five weeks to distill the 10 grams that the Mayo team requested, but by the end of April, a batch of streptomycin was on its way from New Brunswick to Rochester. On April 27, Hinshaw and Feldman began testing it on a dozen different pathogens they successively injected into four very unlucky guinea pigs.

The results were more than encouraging. Streptomycin was effective against nearly all of them: bubonic plague, tularemia, even the intestinal disease known as shigellosis. Most important: By the time the tests were completed on June 20, the preliminary results were almost too good to be true. Streptomycin cured tuberculosis.

At least, it cured the disease in four rodents. To know the compound's real effectiveness, the test would need to be replicated as a proper experiment, with more subjects and a like number of controls: guinea pigs who would be given the disease, but not the treatment. For that second, crucial bit, considerably more streptomycin was needed than Schatz and his Heatley-like factory could dream of producing. On July 9, Hinshaw and Feldman arrived in New Jersey, this time so that Selman Waksman could introduce them to his patrons from Merck at their Rahway lab.

Feldman and Hinshaw had, perhaps overoptimistically, already selected and infected twenty-five guinea pigs before leaving Rochester, but their arguments failed to persuade Merck's team. The pharmaceutical company's chemists and pharmacologists were extremely dubious about their ability to produce the quantities needed for a proper trial. And even if they could, they questioned the wisdom of allocating resources to an unproven drug when the need for more and better penicillin production was clearly a greater national priority. Merck—along with the other sixteen companies that the War Department had enlisted in the penicillin project—was fully committed to the Florey-Chain-Heatley process. Its manufacturing facilities (and, particularly, its fermentation vats, which would be essential for producing large quantities of either penicillin or streptomycin) were working three shifts a day to produce the penicillin demanded by the war effort. This was only sensible, since the most common infections resulting from the wounds to American soldiers, sailors, and airmen were caused by Gram-positive bacteria: *staphylococci, streptococci, enterococci,* and especially *clostridia: C. tetani,* the vector for tetanus toxemia, and *C. perfringens,* which caused gas gangrene. Penicillin killed them. Streptomycin—even if it worked—wouldn't.

At the moment it must have seemed to Waksman, Feldman, and Hinshaw (and even the absent Schatz) that their research was being sidetracked just when they were tantalizingly close to a breakthrough. George Merck joined the meeting. Merck's chief executive knew, better than anyone in the room, the importance of the penicillin project to both the company and the war effort. He also possessed, more than anyone in the room, a vision of the future of drug development. And he alone had the executive authority to decide which was more important. Merck overruled his scientific staff and agreed to invest in a production line for streptomycin in the plant in Rahway; a month later, the company broke ground on a brand-new $3.5 million facility in Elkton, Virginia.

Even better: He directed Randolph Major, Merck's research director, to assign fifty researchers to the project. Major knew precisely who he wanted to head the team. Ten years before, he had hired two Ivy League–trained chemists, Max Tishler and Karl Folkers. He assigned Tishler to work on a series of challenging problems in chemical synthesis—vitamin B_2, cortisone, and ascorbic acid—while Folkers spent two years studying the poisonous alkaloids collectively known as curare as a possible anesthetic.* Both were put on the penicillin project in 1943, and were working on it seven days a week when Major reassigned them to streptomycin. Folkers became the company's head of research, Tishler its head of development.

Tishler and Folkers were both children of immigrant families who received the most prestigious and rigorous graduate education available in the United States—Tishler was awarded his PhD from Harvard's chemistry department, where he studied under the soon-to-be president of the university James Bryant Conant; Folkers did his postdoctoral work at Yale. Both joined Merck in the 1930s precisely because the company's brand-new research lab promised the kind of support available nowhere else. Folkers, in particular, chose Merck over a higher-paying job working for General Electric because his "lab" at GE was more like a storeroom "with some chicken wire to separate it from the rest of the area."

Working together as a team for the first time on streptomycin, Tishler and Folkers got on brilliantly, both with one another and with Waksman, whom Tishler remembered forty years later as an "extremely imaginative, able, wonderful scientist, and a very dedicated and prolific writer. . . . He was probably the best living scientist of the soil; no one has approached his expertise since then." No one was better suited by training and temperament to get a sufficient quantity of streptomycin from New Jersey to the eagerly awaiting Hinshaw and Feldman.

By mid-July 1944, they had received enough streptomycin to begin

---

* Randolph Major put him to work with a Russian botanist named Boris Krukoff, who brought both the curare and a seven-foot-long blowgun from South America to Rahway.

their world-changing experiment. The twenty-five infected guinea pigs were injected with streptomycin every six hours for sixty-one days. Twenty-four control animals were not. The results, even now, are startlingly obvious. After forty-nine days, eighteen of the control animals showed tubercular nodes in their lungs; only one of the treatment animals did. Eight control animals had tuberculosis in their livers; none of the treatment animals did. And seventeen of the untreated animals— 71 percent—had died. Of the twenty-five guinea pigs given streptomycin, twenty-three had survived to the end of the experiment.*

Waksman, Schatz, Hinshaw, Feldman, Folkers, and Tishler had their answer. Another wonder drug, one that seemed to be in every way as transformative as penicillin, had been identified, isolated, and tested.

Who would get to tell the world?

Streptomycin, like penicillin before, and tetracycline and others after, was the work of dozens of highly trained and ambitious professionals, and their motivations aren't easily categorized. For some, it's all about the pleasures of puzzle solving: Howard Florey famously said, "People sometimes think that I and the others worked on penicillin because we were interested in suffering humanity. I don't think it ever crossed our minds about suffering humanity. This was an interesting scientific exercise, and because it was of some use in medicine is very gratifying, but this was not the reason that we started working on it. . . ." For others, though, Florey included, recognition was the thing. Scientists dream of the undying fame that comes with Copley Medals and Nobel Prizes, but even at more modest levels, authoring a major paper is the key to academic status and even employment. This was true even outside of universities; out of a shared conviction, promoted by Major, that Merck needed to be preserved as an attractive place for academics to work, Tishler and Folkers developed

---

* It is, nonetheless, worth noting that all the autopsied guinea pigs still showed evidence of the *M. tuberculosis* bacterium. Streptomycin had prevented disease, but hadn't killed the pathogen causing it; that is, it was *bacteriostatic*, rather than *bactericidal*.

a protocol whereby Merck scientists were allowed—even encouraged—
to publish results any time after a patent was filed, rather than waiting
for the patent to be issued. And, let it not be said that scientists, even
those not employed by profit-making enterprises like Merck, are immune to financial attractions, both from grants and the patent revenues whose loss had so enraged Ernst Chain.

What all these forms of compensation share, though, is the premium they place on priority: Recognition, prizes, grants, and patents
are rewards only for first-place finishers. The problem with the news
about streptomycin was that there were two important discoveries:
first, the Rutgers discovery that the compound produced by *S. griseus*
was effective against Gram-negative bacteria in vitro; and second, the
Mayo results that it worked just as well in vivo. Complicating matters
was the fact that Feldman and Hinshaw, because of their affiliation,
were able to guarantee that their own paper could appear in the *Proceedings of the Mayo Clinic*—a prestigious, peer-reviewed journal—
within weeks of submission, while Schatz and Waksman were still
awaiting a date from the *Proceedings of the Society for Experimental
and Biological Medicine*.

It was a polite disagreement, but a nontrivial one, and it required
a third party to adjudicate it. Waksman persuaded Randolph Major
to—politely—dissuade the Mayo team from publishing until the
Rutgers streptomycin paper could appear. Since Merck had a long-
standing relationship with Waksman (to say nothing of an interest in
controlling the public release of information about a potential new
miracle drug before it was closer to being produced in quantity), Major was happy to tell Feldman that while he "quite understood" his
eagerness to publish, "You might care to wait until the publication of
[Waksman's] initial *in vitro* results."

Major—and, presumably, his boss, George Merck—didn't care
about publishing priority. But he did have a powerful regard for the
company's relationship with Selman Waksman's lab, which was starting to look like one of the most important research investments the
company had ever made. Waksman knew this, and made his case to
his corporate patrons accordingly: He needed to publish first.

It was at roughly the same time that Waksman performed an even more unlikely act of salesmanship: persuading George Merck to give up the patent on streptomycin.

Because of the 1939 agreement, Merck & Co. owned the patent on a potential super antibiotic, while penicillin could be produced by anyone as a generic drug. So the fact that Merck did, in fact, transfer ownership of the patent to the Rutgers Endowment Foundation seems, on the face of it, mystifying. Though George Merck famously, and sincerely, believed that people came before profits, he was also the CEO of a for-profit company that was certainly capable of providing streptomycin to millions of people; and since *someone* was going to own the patent, why not Merck?

The best guess is that, like penicillin (and so much else), the fate of streptomycin was intimately tied up with the overarching strategic objective of the time: winning the Second World War.

Even before it had become a shooting war, the nations that would become the Allied powers had been worrying—probably excessively— about the potential development of biological weapons by the Axis. It was widely believed in London and Washington that Japanese agents had tried to buy yellow fever virus as early as 1939. Other intelligence reports suggested that German biologists were secretly teaching their Japanese partners how to use anthrax and typhus militarily. The Swiss had reportedly told the United Kingdom that Nazi Germany had plans to use "bacteria of every description" in any coming war.

The identities of everyone in the United States who was privy to this intelligence aren't known, but one was certainly George Merck, the head of the War Research Service.

As documented by the journalist and author Peter Pringle, and substantiated by the National Academy of Sciences committee on biological warfare, and the War Research Service, Merck had met with the Office of Strategic Services, a wartime intelligence agency, in 1943 in response to a request that he recommend biological weapons for use in covert operations behind enemy lines. But his more important duty was to protect the Allies against any likely biological attack; and it seems at least plausible that he decided that his responsibility to his nation trumped the one he had as the head of his family corporation. If streptomycin was a potential defense against German or Japanese biological warfare, it was imperative that production be scaled up immediately.* By the end of 1944, he agreed to forgo the monopoly granted by the 1939 agreement with Selman Waksman; within two years, six companies—Abbott, Lilly, Merck, Pfizer, Squibb, and Upjohn—were all manufacturing streptomycin.

Whatever Merck's motives, his decision freed Waksman to secure the patent rights on behalf of Rutgers, and, on February 9, 1945, Schatz and Waksman jointly applied for a patent on streptomycin, swearing under oath that they were codiscoverers of the new drug. Three months before, Merck-manufactured streptomycin had been used, for the first time, on a human being: Patricia Thomas.

Thomas had spent more than a year at Mineral Springs, her symptoms gradually worsening, especially in her right lung. In October 1944, her physician, Dr. Karl H. Pfuetze, sent her to the Mayo Clinic for a consultation and examination by Corwin Hinshaw, who immediately started her on the course of streptomycin that would save her life. Over the next five months, she received five courses comprising multiple doses . . . doses that, in the absence of any prior human testing, were best-guess estimates by her physicians. Mayo Clinic surgeons operated on her nearly useless right lung, excising the diseased section and returning most of its function. In the summer of 1945, she returned home, where she would marry and have three children. Streptomycin saved her life, but she was still permanently weakened by the disease, and would die in June 1966, at only forty-two years old.

For Waksman, 1946 was an *annus triumphalis*, the first of many.

---

* In 1946, the War Department published George Merck's report on Japanese biological warfare research, which concluded, "Intensive efforts were expanded [sic] by Japanese military men toward forging biological agents into practical weapons of offensive warfare [and] pursued with energy and ingenuity."

He traveled to Europe, where he was presented with the first of what would become twenty-two honorary doctoral degrees and sixty-seven prizes and awards, including the Lasker Award, the Trudeau Prize, and the Nobel Prize in Physiology or Medicine. Also, in 1946, on May 3, Waksman and Schatz both assigned their pending patent* to the Rutgers Foundation (by then the Rutgers Research and Endowment Foundation) in return for one dollar apiece.

Schatz, who had by now completed his PhD dissertation, watched from the sidelines as Selman Waksman became a national hero. In 1948, a dozen newspapers hailed Waksman as the discoverer of the new wonder drug; in April 1949, *Time* profiled him as the model of a humble scientist in an article entitled "Man of the Soil." In November, it put him on the magazine's cover, accompanied by the folksy reminder that "the remedies are in our own backyards." Schatz was unmentioned in either article. He registered his concern in a letter to his onetime boss. Waksman's reply deserves quoting in detail:

You know quite well that we gave you all the credit that any student can ever hope to obtain for the contribution that you have made to the discovery of streptomycin. You know quite well that the methods for the isolation of streptomycin had been worked out in our laboratory completely long before your return from the army, namely for streptothricin.

Unsurprisingly, this failed to mollify Schatz. Nor did a subsequent letter in which Waksman told his onetime grad student and co-patent holder:

You must, therefore, be fully aware of the fact that your own share in the solution of the streptomycin problem was only a small one. You were one of many cogs in a great wheel in the study of antibiotics in this laboratory. There were a large number of graduate students

---

* It would be awarded in 1948; significantly, the patent recognized that streptomycin was *not* a naturally occurring substance but a "new composition of matter."

and assistants who helped me in this work; they were my tools, my
hands, if you please.

The "man of the soil" had dug one shovel too deep. In March 1950,
Schatz filed suit in federal court.

More than reputation was at stake. In the intervening years,
Schatz had learned—it's unclear how—that the 1946 assignment of
patent rights to the Rutgers Foundation, for which both Schatz and
Waksman agreed to accept a single dollar in compensation, wasn't the
only document signed by Waksman that month. He had, in addition,
signed an agreement with the foundation that provided him with 20
percent of the royalties they would henceforth receive for licensing
the rights to manufacture streptomycin. Even better (or worse),
Schatz discovered that the agreement between Waksman and the
foundation was contingent on Waksman persuading Schatz to sign
away his rights. For a dollar.

By 1950, Waksman's 20 percent had paid him approximately
$350,000.

Though Waksman attempted a defense against the suit, he was
severely handicapped by the documents he had himself signed—not
merely the patent application, with its solemn oaths that he and
Schatz were codiscoverers, but the original papers in which the discovery of streptomycin was described. Back in 1944, two articles had
appeared in *The Proceedings of the Society for Experimental and
Biological Medicine* entitled, respectively, "Streptomycin: A Substance Exhibiting Antibiotic Activity Against Gram-Positive and
Gram-Negative Bacteria" and "Effect of Streptomycin and Other Antibiotic Substances upon *Mycobacterium tuberculosis* and Related
Organisms." The authors were Albert Schatz, Elizabeth Bugie (another of Waksman's graduate students), and Selman Waksman . . . in
that order. And while Waksman later asserted that it was his policy to
give students first position on papers to help their careers, he did so
only twice in his entire publishing life. Both were the streptomycin
papers he had coauthored with Schatz.

In December, the case was settled without trial. Schatz was granted

3 percent of the foundation's streptomycin royalties, which would amount to about $12,000 a year, and a one-time payment of $125,000 for foreign patent rights. Waksman was granted 10 percent (and 7 percent was granted to everyone else who had worked in the lab during the key months of 1943, up to and including the dishwasher).

If money had been at the heart of the dispute, the settlement should have, well, settled things. Possibly, had the judges who awarded the Nobel Prize in Physiology or Medicine in 1952 recognized Schatz as well as their honoree, Waksman, it might have done so. Or if Waksman had, in his Nobel Lecture, done more than mention Schatz, in a single sentence, as one of twenty different lab assistants. But probably not. To his dying day, Schatz refused to recognize the fact that Waksman was a far more accomplished scientist, one who had made a dozen different landmark discoveries both before and after Schatz's tenure in the lab. In 1949, Waksman even discovered another actinomycete-derived antibiotic, neomycin. Moreover, though Schatz spent decades insisting he was academically marginalized for pursuing on his just rights, it's not hard to see why other biology departments were wary of hiring someone who had attacked his thesis advisor in print, in court, and even, astonishingly, by sending an open letter to the king of Sweden in an attempt to sabotage the Nobel Prize ceremony. In Selman Waksman's mind, on the other hand, the great achievement was entirely the result of an experimental machine he had designed and built decades before Schatz arrived in New Brunswick—and that Albert Schatz was, therefore, an easily replaceable component.

And there lies the real misunderstanding. In a letter written on February 8, 1949, Waksman—again turning up the dial on his indignation—wrote to Schatz, "How dare you now present yourself as innocent of what transpired when you know full well that you had nothing to do with *the practical development of streptomycin*" (emphasis added). Unwittingly, Waksman had revealed the underlying truth of the scientific discovery; and it didn't serve either Schatz or himself. He was correct that Schatz had little to do with the "practical development" of the new antibiotic. But neither had he. William Feldman and Corwin Hinshaw at Mayo had more to do with demonstrating the therapeutic value of streptomycin than anyone at Rutgers. So had Karl Folkers and Max Tishler and the more than fifty researchers assigned by George Merck to supervise the streptomycin project. The discovery of streptomycin was a collective accomplishment dependent on dozens, if not hundreds, of chemists, biologists, soil scientists, and pathologists.

And one statistician.

Streptomycin was a miracle for those suffering from tuberculosis and other bacterial diseases unaffected by penicillin. It was iconic—as much, in its way, as penicillin—in demonstrating the way industrial support could turbocharge university research. But it achieved its most consequential result as the subject of the most important clinical trial in medical history: the one that, for the first time, and forever after, quantified the art of healing.

Though it is frequently described as such, the streptomycin trial of 1948 to 1950 wasn't anything like medicine's first clinical trial. The Bible's Book of Daniel records King Nebuchadnezzar's more or less accidental test of two different diets—one, by royal commandment, only meat; the other vegetarian—over ten days, after which the perceived health of the vegetarians persuaded the king to alter his required diet. The sixteenth-century French surgeon Ambroise Paré tested two different treatments for wounds—one, a noxious-sounding compound made of egg yolks, rose oil, and turpentine; the other of boiling oil, which was even worse. Two centuries later, the Scottish physician James Lind "selected twelve patients [with] the putrid gums, the spots and lassitude, with weakness of the knees" typical of the deficiency disease scurvy, and gave two of them a daily dose of a "quart of cyder," two of them sulfuric acid, two vinegar, two seawater, two a paste of honey. The final two, given oranges and lemons, were marked fit for duty after six days, thus clearly demonstrating how to prevent the disease, a scourge of long maritime voyages.*

---

* Not clearly enough to convince the Royal Navy, though. It took another fifty years before the notoriously conservative Lords of the Admiralty issued an order that

Even fictional characters got into the act. The climax of Sinclair Lewis's 1925 novel, *Arrowsmith*, takes place on the fictional Caribbean island of St. Hubert's, in which the eponymous hero gives half the population of the parish of St. Swithin's a vaguely described anti-plague serum—a "phage"—and the other half nothing at all.

Nor was the streptomycin trial the first time medicine had recognized the importance of sampling. The *idea* of sampling—choosing a test population so that it reflects the characteristics of the entire universe of similar people, and large enough that conclusions wouldn't be confounded by a tiny number of outliers—was, in some sense, centuries old; the sixteenth-century Dutch physician and chemist Jan Baptist van Helmont had dared his fellow doctors to compare their techniques with his, proposing:

> *Let us take out of the hospitals . . . 200 or 500 poor People that have Fevers, Pleurisies, etc. Let us divide them into halfes, let us cast lots, that one of them may fall to my share, and the other to yours [and] we shall see how many funerals both of us shall have. . . .*

William Farr, compiler of abstracts in Britain's General Register Office, first identified the huge difference in childhood mortality between rich and poor, leading to thought experiments on sanitation. And the pioneers of mathematical statistics, Karl Pearson and Ronald Fisher, had applied techniques like analysis of variance and regression to a variety of health-related subjects, such as height and blood pressure. In 1943, Britain's Medical Research Council actually funded a comparative trial to see if an extract of *Penicillium patulinum* could cure the common cold (it couldn't). During the first four decades of the twentieth century, more than one hundred papers cited so-called alternate allocation studies, in which every other patient admitted to a hospital or clinic was given a treatment, and the others were

---

made drinking lemon, and later lime, juice—hence "limeys"—compulsory for sailors on long voyages.

used as a control. But before streptomycin, the perceived efficacy of any medical treatment remained entirely anecdotal: the accumulated experience of clinicians.

This is the sort of thing that works just fine to identify really dramatic innovations, like the smallpox vaccine, or the sulfanilamides, and certainly penicillin. But most advances are incremental; this is true not just for medical innovations, but all technological progress. And identifying small increments of improvement isn't a job for clinicians. Physicians, no matter how well trained in treating disease, are just as vulnerable as anyone else to cognitive biases: those hiccups of irrationality that give an excessive level of significance to the first bit of information we acquire, or the most recent, or, most common of all, the one we hope will be true.* Medicine needs statisticians.

It was a lucky coincidence that the most influential medical statistician of the twentieth century had a special interest in tuberculosis. Austin Bradford Hill, professor of medical statistics at the London School of Hygiene and Tropical Medicine—like Selman Waksman, a nonphysician—had spent the First World War as a Royal Navy pilot in the Mediterranean, where, in 1918, he acquired pulmonary tuberculosis. Treated with bed rest and the dangerous, unproven, but nonetheless popular technique of deliberately collapsing his lung, he somehow recovered, and was granted a full disability pension, which he would collect for the next seventy-four years, until his death in 1992.

Hill's personal connection with tuberculosis wasn't the only reason that streptomycin was the ideal candidate for the then-revolutionary idea of randomizing a population of patients and carefully comparing outcomes between those who received a particular treatment and those who didn't. Because tuberculosis was so variable in its symptomology—most people who have *M. tuberculosis* are asymptomatic; some will have a chronic ailment for years; and others die within weeks—it was, even more than most diseases, subject to the "most people get well" problem.

---

* In the taxonomy of cognitive science, these are known, respectively, as *anchoring*, the *availability heuristic*, and *confirmation bias*.

And, as all those people who flocked to sanatoria proved, tuberculosis was highly susceptible to environmental differences: People really *did* do better at Saranac Lake than in cities, where clean air and water were still rare. Streptomycin didn't cure tuberculosis the way penicillin cured staph infections; before Patricia Thomas left Mineral Springs, she received five courses of treatment over more than four months. Teasing out the best practices for a treatment that worked that slowly—slower than some diseases resolve in the absence of any treatment at all—*really* demanded sophisticated mathematics.

Despite the obvious need for statistical analysis of tuberculosis treatments, Bradford Hill had another obstacle to overcome in persuading Britain's Tuberculosis Trials Committee, which he had joined in 1946, to fund a randomized trial to evaluate streptomycin: ethics.

Such a trial would mean denying what was widely (and correctly) believed to be a lifesaving drug to the members of a control group. But without a control group, it would be impossible to arrive at a definitive conclusion about the treatment group. This was an especially thorny issue in 1946, as the Nuremberg Military Tribunals were, at that very moment, revealing the horrific results of human experimentation in Nazi Germany, where patients were frequently denied treatment in the name of science.* Letting chance decide which tuberculosis patients got the miracle drug, and which would serve as a control group, would probably have been an insuperable problem, but for one thing: There just wasn't enough of the drug to go around. A significant number of tuberculosis patients would be denied treatment in any case. As Hill subsequently wrote:

---

* Not everyone was as punctilious about ethics as the Tuberculosis Trials Committee. At the moment in 1947 when the war crimes tribunal convicted seventeen of the twenty-three defendants in the so-called Doctors' Trial and published the ten-point Nuremberg Code for human experimentation, the infamous Tuskegee syphilis experiments had already been under way for fifteen years, and would continue observing the course of untreated syphilis in African American men for another twenty-five years.

We had exhausted our supply of dollars in the war and the Treasury was adamant we could only have a very small amount of streptomycin . . . in this situation it would not be immoral to do a trial—it would be immoral not to, since the opportunity would never come again. . . .

The Medical Research Council's Trial of Streptomycin for Pulmonary Tuberculosis began early in 1947 with 107 patients in three London hospitals. All of them were young, and had recently acquired severe tuberculosis in both lungs. Fifty-five of them were assigned to the treatment group, which would be given streptomycin and bed rest, while the fifty-two members of the control group would receive bed rest and a placebo: a neutral pill or injection indistinguishable to the patient from the compound being tested. The two groups were selected entirely at random, based on numbers chosen blindly by the investigator and placed in sealed envelopes, thus assuring that no selection bias would occur unconsciously. Nor were the patients themselves told whether they were receiving streptomycin or a placebo.

Hill insisted that results of the test wouldn't depend on clinical evaluation alone, but on changes in chest X-rays, which would be examined by radiologists unaware of whether the subject had been in the treatment or the control group. The exercise was therefore a real beauty: a randomized, triple-blind (neither patients, nor clinicians, nor evaluators were told who had received the treatment) clinical trial, the first in the history of medicine.

The results were just as compelling as the methodology, though in an unexpected way. After six months, twenty-eight patients on the streptomycin regimen had improved, and only four had died. The control group, meanwhile, had lost fourteen of its fifty-two patients.

But while streptomycin "worked," the rigor that Hill had built into the experiment's design revealed streptomycin's weaknesses just as clearly as its strengths, even for treating tuberculosis. Because it took months for the treatment to show a statistically demonstrable effect, some of the bacteria were certain to develop resistance to the therapy *while it was still going on*. And, indeed, when the subjects were revisited three years later, thirty-five of the members of the control group had died . . . but so had thirty-two who had received the treatment.

The experiment threw a massive dose of reality-based cold water on the enthusiasm of the first clinical reports. Something more than streptomycin was clearly required to redeem the drug's initial promise. Luckily, Hill had a good idea what the "something more" should be. In 1940, a Danish physician, Jörgen Lehmann, had reasoned that, since acetyl salicylic acid, the compound better known as aspirin, increased the oxygen consumption of *M. tuberculosis*, its chemical cousin—para-aminosalicylic acid, or PAS—might act to inhibit oxygen consumption, thus killing (or at least disabling) the aerobic bacterium. It was a decent enough theory, and in 1946 Lehmann had published an article in the *Lancet* with his results, which were modest enough. By itself, PAS was only a marginally effective treatment for tuberculosis. But because its mechanism worked to inhibit oxygen consumption by the bacterium, the logic went, it strengthened the action of streptomycin, which needs oxygen to enter the bacterial cells.*

In his second trial, beginning in December 1948, Hill duplicated exactly the same experimental structure—same randomization, same X-ray evaluation—as his first. This time, however, he added Lehmann's oxygen inhibitor to the treatment group. Less than a year later, the power of what would come to be known as RCT, for randomized controlled trials, was vindicated. The Medical Research Council announced that they had shown "unequivocally that the combination of PAS with streptomycin considerably reduces the risk of the development of streptomycin-resistant strains of tubercle bacilli." The three-

---

* Or so they thought. Later researchers have found a more plausible mechanism for the way PAS improved the effectiveness of streptomycin: The former drug slows down the bacterium's ability to produce folate, while streptomycin inhibits RNA synthesis, thus exposing the bacterium to what is, effectively, a two-front war.

year survival rate using the combination of the two drugs was an almost unbelievable 80 percent.*

In less than three years, penicillin and streptomycin had achieved more victories in the battle against infectious disease than anything in the entire history of medicine since Galen. Both were unprecedentedly powerful weapons against pathogens; but it was streptomycin that revealed a method for finding more of the same: the combination of Selman Waksman's protocol for finding antibacterial needles in haystacks made of soil, and Bradford Hill's arithmetic for revealing their clinical value.

* The PAS-streptomycin cocktail was also the precursor of the so-called fixed-dose combination antibiotics that would be the target of the most far-reaching federal investigations of the drug industry since the Massengill debacle of 1938. See Chapter Nine.

SEVEN

"Satans into Seraphs"

The perennial known as timothy grass, which grows from two to five feet tall, covers thousands of acres of the American Midwest. It is famously hardy, resistant to both cold and drought, and prospers in almost any kind of soil, from the heaviest bottomland to the poorest sands. Like many New World plants, it is a relatively recent invader, introduced to colonial America by European settlers—one popular theory suggests that the name comes from an eighteenth-century New England farmer named Timothy Hanson—and it is widely grown as animal feed for everything from domestic rabbits to cattle and horses.

Timothy grass was and is important enough as a commercial commodity that agronomists at the University of Missouri started planting it in Sanborn Field, the university's agricultural test station, as soon as it opened in 1888. They were still experimenting with it—testing varieties for improved yields, or more weather hardiness—in 1945 when William Albrecht, a soil microbiologist, received a letter from a former colleague, now working in New York. The letter included a request that Albrecht obtain soil samples from a dozen different Missouri locations, including Sanborn Field's Plot 23. Its author was a botanist and mycologist named Benjamin Minge Duggar.

Duggar was then seventy-three years old and an accomplished and respected plant pathologist. Ever since receiving his PhD in 1898, he had studied fungi and disease, more or less nonstop, at the Department of Agriculture and at a number of prominent land-grant universities including Cornell, the University of Wisconsin, Washington University in St. Louis, and, of course, Missouri.* In 1944, he departed his last academic post and joined Lederle Laboratories to work under its remarkable head of research, Dr. Yellapragada Subbarao.

Lederle Antitoxin Laboratories, as it was originally known, had been founded in 1904 by Dr. Ernst Lederle, a former New York City health commissioner, to produce an American version of the diphtheria vaccine developed by Emil Behring, Paul Ehrlich, and Robert Koch at the end of the nineteenth century, one they could sell to American physicians and hospitals royalty free. Vaccines and antitoxins, for tetanus, typhoid, anthrax, and smallpox, remained the company's primary business for the next forty years, through the death of its founder in 1921, its subsequent acquisition by the agricultural chemicals manufacturer American Cyanamid in 1930, and the hiring of Subbarao in 1940.

Subbarao, an Indian-born physician and physiologist, arrived in the United States as a penniless immigrant in 1923, but with an admissions letter to Harvard University's School of Tropical Medicine, a division of the university's medical school. His tuition expenses were paid by his father-in-law, but in order to pay for his room and board he was given a job at Harvard Medical School, where he spent the next seventeen years. His achievements were nothing short of stellar; among other endeavors, he isolated the components of adenosine triphosphate, or ATP, the fuel for all cellular respiration. In fact, a complete account of Yellapragada Subbarao's accomplishments is almost literally too long to list—not only fundamental discoveries about ATP, creatine, and of B_{12}, but half a dozen chemical breakthroughs still in use today, including discovering how a mimic of

---

* The land-grant colleges were, like the Agriculture Department itself, a creation of the 37th United States Congress, which passed the Morrill Act—named for its author, Vermont congressman Justin Smith Morrill—in July 1862 to "teach such branches of learning as are related to agriculture and the mechanic arts. . . ." In the same way as the Northern Lab and Selman Waksman's lab at Rutgers, they are a reminder of the unpredictable but enormous returns earned by national investment in basic research.

folic acid known as antifolate could be used to combat leukemia. Despite that, U.S. immigration law's baroqueries (among other things, immigrants from British India were allowed to stay only if they fell into professional categories that the State Department deemed valuable . . . a list that changed regularly) required him to register as an alien for his entire professional career.

The most consequential result of Subbarao's problematic immigration status is that one of the university's most brilliant investigators was denied tenure. Academia's loss was industry's gain; in 1940, he left to join Lederle as its director of research. A year after that, he represented Lederle at the first meetings of the Committee on Medical Research called by A. N. Richards to discuss what would become the penicillin project. Three years after that, he hired Benjamin Duggar.

By this time, Selman Waksman's researches at Rutgers were making him the most famous soil scientist in the world; more, they were inspiring everyone in the entire discipline to emulate his approach: testing literally thousands of actinomycetes for antibacterial properties. It certainly inspired Subbarao and Duggar, who initiated a global program of soil collection. Remarkably enough, in the middle of the world's largest war, they successfully recruited dozens of soldiers and sailors to seek out soil samples everywhere from the Caucasus to North Africa to South America.

In 1945, they hit pay dirt—literally—closer to home, with the sample from Plot 23 at Sanborn Field. Living in Plot 23 was a yellow actinomycete, a relative of Selman Waksman's *Streptomyces griseus* that they designated A-377. It took nearly three years of testing and experiment before Duggar announced his discovery to the world in an article in which he named his newly discovered organism *Streptomyces aureofaciens*: the "gold maker."

The name was almost certainly an attempt to describe the appearance of the bacterium, but Duggar's employers at Lederle may have had a different meaning in mind. *S. aureofaciens* produced a chemical of unknown identity and structure, but one that checked the activity of an enormously wider variety of bacteria than either penicillin or streptomycin. The substance, which Duggar christened

Aureomycin (*aurum* is Latin for gold), was effective against Gram-positive and Gram-negative bacteria, including the pathogens responsible for common ailments like urinary tract infections, and rare ones like bubonic plague. It even seemed to have a powerful effect against a number of viruses. The first broad-spectrum antibiotic had been discovered.

In 1948, after a series of highly successful animal experiments, Aureomycin was ready for clinical investigation in humans. The facility chosen was Harlem Hospital, where Louis Tompkins Wright had spent years studying the treatment of diseases, such as the sexually transmitted lymphatic infection caused by the bacterium known as *Chlamydia trachomitis*. Wright—the most famous African American physician in the United States, the first admitted to the American College of Surgeons—succeeded brilliantly, not just against chlamydia, but also on the variety of pneumonia caused by a virus, rather than the *pneumococci* bacteria.

Aureomycin looked like a true magic bullet: the hoped-for drug that would cure nearly everything. The Harlem Hospital results didn't convince everyone; Maxwell Finland of Harvard Medical School, perhaps the country's most respected expert on infectious disease, found that Wright's reports were "tinged with enthusiasm," and he didn't mean it as a compliment. In 1948, this made him a voice in the wilderness.* Lederle promoted it as "the most versatile antibiotic yet discovered, with a wider range of activity than any other known remedy." It wasn't just that it was superior to existing antibacterial treatments in treating disease (though it was). Unlike Prontosil, or penicillin, or—thanks to George Merck—even streptomycin, Aureomycin was patentable, and, on September 13, 1949, it was granted U.S. patent number 2,482,055. Even before the patent had been approved, in 1948 the company spent a then-unheard-of $2 million shipping samples of their gold maker to 142,000 doctors. Lederle had the first blockbuster drug in history, but it wouldn't have the stage to itself for long.

* For more on Max Finland, see Chapter Nine.

In 1945, Brooklyn-based Pfizer had initiated its own global soil collection program with the same goal as Lederle: to find a patentable antibiotic. They went about it the same way that Waksman and Schatz had discovered streptomycin, by testing large numbers of soil samples—*very* large numbers. Within a few years, Pfizer had collected more than 135,000 of them. One of Pfizer's chemists, Ben Sobin, later recalled, "We got soil samples from cemeteries; we had balloons up in the air [that] collected soil samples that were windborne; we got soil from the bottoms of mine shafts . . . from the bottom of the ocean." By the beginning of 1949, its investigators had conducted more than twenty million tests on them at the company's Terre Haute, Indiana, facility, a state-of-the-art microbiology laboratory.

In a replay of Lederle's experience, the payoff came not from an exotic location but from the land around one of Pfizer's own Midwest plants. They had found a yellowish actinomycete that Alexander Finlay, Pfizer's team leader, named *Streptomyces rimosus*.* The canary yellow crystals it produced were given a code name: PA-76, for the seventy-sixth culture of a Pfizer antibiotic.

At first, PA-76 seemed to be virtually identical to Lederle's Aureomycin, which made it interesting but commercially problematic. Pfizer nonetheless proceeded to invest, since PA-76 had enormous potential to become Pfizer's own blockbuster. It killed or at least slowed down Gram-positive and Gram-negative bacteria as well as dozens of fungi and, seemingly, even viruses. The honor of naming it was given not to its discoverer, but Pfizer's new president, John McKeen. McKeen had designed and converted Pfizer's Brooklyn fermentation plant, a former ice factory, for the wartime penicillin project. He had risen through the ranks to succeed John L. Smith as Pfizer's president, and named it Terramycin, because, as he later wrote, "I wanted a name connected with the earth, and one that could easily be recalled by doctors and scientists and people in general, because it came from the earth."

---

* *Rimosus* translates roughly from the Latin as "leaking" or "full of holes," which is vivid, though less metaphorically resonant than "gold maker."

In 1949, McKeen assigned Gladys Hobby, the microbiologist who had been part of the Columbia University team that led the way in American experimentation on penicillin, now a Pfizer team member, to take charge of testing the new drug. She wasted little time. On December 31, 1950, Hobby conducted Terramycin's first human trial at Harlem Hospital, which had demonstrated in the Aureomycin experiments that it had the personnel and structure in place (including carefully identified pathogens) to perform a high-quality clinical trial.

Her employers couldn't have been more enthusiastic about her results. Though Pfizer had produced huge quantities of penicillin and streptomycin—far more than any other pharmaceutical company in the world—they weren't making much money selling it. In March 1950, McKeen famously gave a speech to the New York Society of Security Analysts, in which he told them, "If you want to lose your shirt in a hurry, start making penicillin and streptomycin." The surest way out of the trap that had swallowed up the first antibiotics was finding a new drug that was superior to its competition, so the company that found it could profit from a de facto monopoly.

Now Pfizer had their drug: Terramycin was just as good as hoped. However, no one at Pfizer or elsewhere thought it was demonstrably superior to Aureomycin, and the Lederle drug already had a head start in winning the hearts and minds of America's physicians and pharmacists. By the beginning of 1950, Aureomycin accounted for 26 percent of the entire antibiotic market in the United States.

Moreover, Pfizer wasn't even really in the business of selling drugs. All of the penicillin and streptomycin they had produced to date had been sold under the label of other companies, ones that had consumer sales and marketing capabilities that Pfizer—which had, after all, been largely a manufacturer of citric acid before joining the penicillin project—lacked. McKeen was undaunted. Terramycin would be a company changer: Pfizer's first branded drug.

But first, the company needed to know more about both Terramycin's mechanism and its structure, a task requiring the most sophisticated understanding of organic chemistry. Pfizer needed the best organic chemist in the world. They needed Robert Burns Woodward.

Woodward was a rare bird: a prodigy—he entered the Massachusetts Institute of Technology in 1933 at age sixteen, and left four years later as a twenty-year-old PhD in chemistry—whose adult achievements were even more prodigious. During his forty-one years at Harvard's Department of Chemistry, he authored or coauthored nearly two hundred peer-reviewed papers, received twenty-four honorary degrees and twenty-six medals and awards, including the National Medal of Science, the Royal Society's Copley *and* Davy Medals, and the 1965 Nobel Prize in Chemistry for his "outstanding achievements in the art of organic synthesis."

Woodward was the first to synthesize cortisone, cholesterol, strychnine, and chlorophyll. In 1944, as an advisor to the War Production Board, he discovered how to synthesize the antimalarial compound quinine, which was utterly essential for the war effort in both southern Europe and throughout the campaign in the Pacific . . . especially so since, from 1941 forward, the world's entire supply of the only natural source for quinine, cinchona tree bark, was under the control of the Imperial Japanese Army. And, seven years *after* winning the Nobel, Woodward succeeded in one of the most impressive tasks in the then-brief history of chemical synthesis, leading the international team that spent twelve years decoding and producing the notoriously complicated molecule known as vitamin B_{12}.

But calling Woodward a master of chemical synthesis, though true, understates his gifts. To his contemporaries, he was even more brilliant at *describing* complex organic chemicals—deciphering the incredibly complicated three-dimensional shapes adopted by the stuff of life—than in *making* them.* In January 1945, it was Woodward who demonstrated that the beta-lactam structure of penicillin, the

---

* Though not, perhaps at first. Max Tishler, Woodward's colleague at Harvard before departing for Merck, later recalled, "None of us thought he was really that great when he came over from MIT. He had such a great press. Of course, we changed our minds pretty quickly."

one proposed by Ernst Chain and E. P. Abraham, had to be correct, thus anticipating Dorothy Crowfoot Hodgkin's X-ray crystallography by five months. In the words of one of his biographers, he could integrate facts "both clear and misleading into a coherent whole better than any chemist who ever lived."

Robert Burns Woodward, 1917–1979

Credit: Getty Images/Keystone

Readers living in a time when technologies like nuclear magnetic resonance imaging are routinely able to determine molecular structure at a relatively low cost in a single afternoon may find it difficult to understand the value placed on Woodward's talent during the 1940s and 1950s. Instead of simply examining a three-dimensional picture of a complicated organic molecule on a screen, chemists of his era could only derive structure by working as enormously sophisticated puzzle solvers: taking all the known facts about a molecule, such as whether, and how quickly, it reacted to heat or cold, to acids or bases, or to other molecules; and from that information, and a detailed knowledge of the laws of chemistry, figuring out which atoms connected to one another, through what sort of bonds, and in what configuration. This is a little like drawing the blueprint of an office building knowing only its floor-by-floor heating bills and the number of people who used its elevators daily.

At this, at the manipulation of what chemistry students know as stereoisomers—alternative spatial configurations of three-dimensional molecules—Woodward was unmatched, not merely because of what might be called his architectural eye, but his profound understanding of the underlying physics of the molecules in question. So, when John McKeen went looking for the chemist best equipped to aid Pfizer in understanding Terramycin, he didn't have to look far.

Nor did he have to do much persuading. Any qualms about working for a commercial employer, which had earlier stymied A. N. Richards during his imbroglio with the American Society of Pharmacology, had disappeared, seemingly overnight. Woodward wasn't ever going to be an industrial chemist per se, but he was more than happy to consult with industry. (One of his best friends was Edwin Land, the founder of Polaroid, who started paying Woodward as a consultant as early as 1942, and eventually made him the only nonemployee to be given options on Polaroid stock, and thereby made him a very wealthy man.) Pfizer had a puzzle that needed solving, and Woodward, who famously completed the *New York Times* crossword daily, needed puzzles to solve.

After dozens of other chemists had tried and failed to figure out the compound's molecular blueprint, Woodward, in legend anyway, took a large piece of cardboard, wrote on it every fact known about the compound, and "by thought alone, deduced the correct structure for Terramycin."

What he reported might have seemed, at first, problematic. Pfizer's new drug was, indeed, not just functionally similar to Aureomycin. It was structurally similar as well. Both compounds were built around a four-ring structure, which gave them the generic name "tetracyclines." But Aureomycin had a single chlorine atom—generically chlortetracycline—that Terramycin lacked. Meanwhile, Terramycin (or oxytetracycline) had an oxygen atom that was missing from Aureomycin. From a medical standpoint, the differences were negligible. But as intellectual property, the discovery was huge. Terramycin was novel enough to be patented. Pfizer filed a patent for Terramycin in November 1949; the FDA approved it five months later. Pfizer was ready to start its production line.

And this time, they wouldn't just be producing a drug for other companies. They decided to sell it themselves.

This was, to put the nicest possible construction on it, ambitious. In 1950, Pfizer's sales force, including the sales manager, numbered only eight people. Undaunted, McKeen sent telegrams to eight hundred drug wholesalers—essentially, every wholesaler in the United States—announcing Terramycin's availability as soon as it was cleared for sale by the FDA. And, on March 23, 1950, within an hour of receiving the formal approval, each of Pfizer's eight sales representatives was manning a switchboard, telephoning a hundred wholesalers each, offering a very steep introductory discount.

Within a year, Pfizer was employing a hundred salesmen. In 1951, the company supplemented them with seventy third-year medical students, hired to work for the summer, and sent them to forty cities to sing the praises of Terramycin. By 1952, three hundred Pfizer reps were selling Terramycin as fast as the company could manufacture it.

Pfizer's efforts—and success—were matched by the tactics Lederle was simultaneously employing with Aureomycin. In the battle for America's antibiotic dollar, featuring ever more elaborate sampling campaigns, twenty-four-hour telephone blitzes (and competition to see which sales reps could drain their respective employer's travel and entertainment budget faster, by entertaining more lavishly), Pfizer and Lederle both emerged victorious, equipped with an arsenal of anti-infective drugs that exhibited a new, and far wider, range of effectiveness.

The biggest reason was that the broad-spectrum antibiotics really *were* new and improved. Penicillin works by weakening the molecules that form the bacterial walls of Gram-positive pathogens; streptomycin disrupts the way that bacteria make protein, though with some nasty side effects, including kidney damage and deafness. Tetracyclines also fight pathogens by inhibiting protein synthesis, but far more effectively: Both Aureomycin and Terramycin hijack the system that bacteria use to accumulate needed molecules from their environment, and so concentrate the bactericidal molecules precisely where they can do the most good, and fast; tetracyclines can accumulate in concentrations more than fifty times greater inside a bacterium than outside.* This made the new drugs effective against virtually every sort of pathogenic bacterium, from the spirochetes that cause syphilis to the bacilli responsible for anthrax and bubonic plague. Moreover, because protein synthesis is such a universal requirement of life, the tetracyclines were also useful against the pathogens that cause malaria—not bacteria, but protozoans, single-celled organisms with nuclei, which evolved billions of years after the first bacteria appeared.

Largely because they generated greater revenues, which supported ever more aggressive marketing, broad-spectrum antibiotics completely overtook penicillin and streptomycin in sales.† By 1952, Americans

---

* For readers who want more detail only: Prokaryotes like bacteria produce the proteins needed for life using a different kind of ribosome than their eukaryotic descendants: everything from fungi to plants to readers of popular history books. The bacterial ribosome, with a different assortment of nucleotides, is broken into subunits, one smaller than the other. Streptomycin binds to the small bacterial subunit, and when it does so, the ribosome misreads the recipe for synthesizing proteins. This makes streptomycin deadly to them, and at least tolerated by humans. The tetracyclines, on the other hand, are far less toxic, because the patient's eukaryotic cells don't absorb them in anything like the same concentrations. This doesn't mean they are completely free of side effects, as detailed below.

† It's less clear that they were clinically superior. Though broad-spectrum antibiotics are effective against a wider range of pathogens, for most infections, penicillin and its synthetic analogues remain the treatment of choice to this day.

were spending more than $100 million annually for broad-spectrum antibiotics, more than three times as much as they were spending on penicillin. The profit margins for the former ranged from 35 to 50 percent, while the profit on penicillin and streptomycin barely topped 5 percent. Pfizer accounted for 26 percent of all antibiotic sales; Lederle a bit more than 23 percent. And the pie was growing larger every year. By the early 1950s, more money was being spent on antibiotics than on all the new and improved patent medicines, toothpastes, mouthwashes, vitamins, hormones, botanicals, and even sulfanilamides, combined.

It took some time before the complicated effects of broad-spectrum antibiotics like the tetracyclines were understood; in some respects, they're not fully grasped even today. At their introduction, the results seemed simple enough: They were true wonder drugs. Unfortunately, though Aureomycin and Terramycin were distinctive enough to have "exclusive" patents, they weren't truly different in any other important way. By late 1952, in fact, Woodward had demonstrated that the molecule that was doing all the heavy lifting for Terramycin wasn't oxytetracycline, but simple tetracycline; inferentially, this meant that Aureomycin, or chlortetracycline, didn't need its chlorine atom, either. It could have been Prontosil all over again: a discovery that the active ingredient in a branded medication was far simpler than the one described in the patent that protected it from competition.

The difference was that sulfanilamide had been discovered so long before it revealed its antibacterial properties that Bayer couldn't protect it as a piece of intellectual property. Simple tetracycline, on the other hand, was new. In October 1952, the Pfizer team filed a patent application for the four-ring molecule at the heart of Terramycin: tetracycline itself. A few months later, in March 1953, Lederle's parent, American Cyanamid, filed its own application. And, just to keep things interesting, in September both Bristol Laboratories and Heyden Chemical—members of the penicillin consortium that had independently discovered how to make tetracycline without starting from either chlortetracycline *or* oxytetracycline—filed patent applications both for the tetracycline molecule and for their own distinctive methods of producing it.

This complicated aspect of the patent system—two molecules that differed from one another by the placement of a single atom could be separately patented, even if the atom was unnecessary to the therapeutic activity of the molecule; two different processes by which identical molecules were produced could, likewise, receive separate patent protections—made for a monumental mess. Though Pfizer was attempting to secure patent protection for simple tetracycline, they had no method for making it that didn't require starting with the chlortetracycline produced by *S. aureofaciens*. This meant that producing a Pfizer-branded tetracycline—they named it Tetracyn— could only be done by licensing (or buying) Aureomycin from Lederle. Lederle, on the other hand, could produce tetracycline only by getting a process license from Bristol or Heyden before they could sell *their* version of plain vanilla tetracycline, which they called Achromycin. Since Lederle planned to launch Achromycin with a $2.5 million campaign—$1 million for samples alone—that included promoting it at two hundred separate professional meetings and sending more than a hundred mailings to every physician (and seven more to every dentist) in the country, some resolution of the license impasse was urgently needed.

Meanwhile, Bristol Laboratories, which was awaiting patent approval for its own method of making tetracycline, would still need a license for the product patent. When they couldn't get one, Bristol initiated a lawsuit against Pfizer, whose own patent application was looking cloudier by the day, as they had apparently withheld information about the original discovery that would have damaged their claim.

The only solution that worked for everyone was a complicated roundelay of cross-licenses. American Cyanamid acquired Heyden's pending patent application for a method of producing tetracycline, then withdrew its own application for a patent on the tetracycline molecule itself. In return, Pfizer granted American Cyanamid/Lederle a license to manufacture the drug. Finally, on the theory that a competitor makes less trouble when its attorneys are inside the boardroom sending cease-and-desist letters *out* than outside the boardroom sending cease-and-desist letters *in*, Pfizer and American Cyanamid agreed to grant cross-licenses to Bristol Laboratories as well: They would be allowed to continue making tetracycline, but to supply only Squibb and Upjohn.

The peace treaty worked brilliantly at achieving its intended objective: propping up the price of the various tetracycline antibiotics, which had fallen by two-thirds between 1948 and 1952, but stabilized thereafter for another decade, mostly for sales of the simplest, most generic form. In 1951, Lederle's Aureomycin was 41.5 percent of the broad-spectrum market; five years later, it was barely 12 percent; but Achromycin, its version of tetracycline, represented 66 percent of sales, and virtually all of the company's $43 million in profit. And they weren't the only ones making money. Squibb sold the antibiotic as Steclin; Upjohn had Panmycin. Even Bristol Laboratories, once they eased out of the restrictions of the original cross-license, sold tetracycline under the name Polycycline.*

Though selling what were, chemically, identical drugs, the companies didn't stop competing. But since they were no longer competing to supply a superior product, and had agreed not to compete on price, victory would go to the best drug marketer. And it is in no way a criticism of Pfizer's research and production brilliance to say that, when it came to marketing, they were truly in a class by themselves.

Their teacher was the legendary advertising executive Dr. Arthur M. Sackler.

For a man whose critics recall him as one of the twentieth century's great hucksters,[†] Sackler's academic credentials are certainly impressive

---

* To the Federal Trade Commission, this looked an awful lot like a conspiracy in restraint of trade, and therefore a violation of the Sherman Antitrust Act. In 1958, the three corporations and their chief executives—American Cyanamid's Wilbur G. Malcolm, Pfizer's John McKeen, and Bristol-Myers's Frederic Schwartz—were sued for conspiring to limit competition and fix prices. The litigation would drag on for twenty-four years before the FTC lost its final appeal.

† Dr. Sackler is even better remembered as one of the great art collectors of the twen-

enough. He graduated from New York University's School of Medicine in 1937 and founded the Laboratories for Therapeutic Research in 1938, while simultaneously completing his residency in psychiatry at Creedmoor State Hospital in Queens. Over the years, he would publish more than 150 research papers, largely in the most rarefied realms of neuroendocrinology and the metabolic basis of schizophrenia. His outsized place in the history of medicine, however, is a result of his mastery of a different aspect of human behavior. In 1942, he joined the William Douglas McAdams advertising agency, which he would, shortly thereafter, acquire.

The timely encounter between a brilliant and ambitious physician, a newly sophisticated advertising business, and a seemingly inexhaustible supply of miracle drugs changed medicine forever. Because Pfizer lacked the long-standing relationships with doctors and hospitals of its competitors, Sackler proposed the company persuade physicians to try Terramycin and Tetracyn not simply at one-on-one sales calls, but through their trade journal: the *Journal of the American Medical Association*. His strategy was clear enough, at least in retrospect. Before 1952, advertising in *JAMA* was almost entirely free of branded drugs. Readers of America's premier physician's journal were far more likely to be introduced to ads for generic medical supplies ("Doctors Get a Heap of Comfort from Grinnell Gloves") and, of course, ubiquitous ads for cigarettes: "More Doctors Smoke Camels Than Any Other Cigarette." Even with that level of support, advertising in *JAMA* was still, by the standards of midcentury magazine publishing, sparse. Drug companies—"ethical" drug companies, as they were still known—hadn't quite erased the line separating medicine from marketing.

By 1955, though, the journal was carrying more advertising pages annually than *Life* magazine, and the number of pages with branded ads increased by a whopping 500 percent, almost all of it from Sackler's

tieth century, whose bequests appear in galleries at Princeton University, Harvard, the Brooklyn Museum, the Smithsonian, and, most prominently, the Sackler Wing at the Metropolitan Museum of Art.

number one client. From 1952 on, Pfizer purchased more than two-thirds of all the antibiotic advertising in *JAMA*. And, if that weren't enough, between 1952 and 1956 virtually every issue of *JAMA* arrived in the offices of hundreds of thousands of physicians with Pfizer's house newsletter, *Spectrum*, bound in. The ads were ubiquitous, and striking. One resembled an eye chart:

O

CU

LAR

INFEC

TIONS

RESPOND

TO BROAD

SPECTRUM

TERRAMYCIN

Eye infections, respiratory ailments, skin lesions: Terramycin treated them all. As part of the strategy of positioning the drug as the antibiotic of choice for the maximum number of potential users, the company even produced a cherry-flavored suspension for children, Pfizer promising to "Turn Satans into Seraphs."

As the various tetracycline producers battled over market share, the impulse to exploit every possible application for their wonder molecules found an unexpected niche.

Almost all of the pharmaceutical firms that were jump-started by the antibiotic revolution had been in the business of manufacturing vitamins since at least the 1930s. Millions of dollars had flowed to the Wisconsin Alumni Research Foundation to license the technology for supplementing milk with vitamin D by irradiating it with ultraviolet light; the New Zealand company Glaxo had grown to be the preeminent supplier of vitamins in the United Kingdom. Ten percent of Merck's sales in 1940 was from vitamins, and they were a significant source of revenue for both I. G. Farben in Germany and Hoffmann-La Roche in France. The discovery that vitamin deficiencies were at the root of dozens of human diseases, from scurvy to rickets, created an enthusiasm for a vitamin-fortified diet that continues to this day.*

One of the most dangerous such diseases had been discovered and named even before vitamins themselves. For nearly a century, medicine had recognized the condition known as pernicious anemia, which is caused by a lack of what had been called, since the middle of the nineteenth century, the "intrinsic factor" needed for production of red blood cells. The disease, sometimes known as Addison's or Biermer's anemia (for the English and German doctors who first identified it), was treatable by eating large amounts of calves' liver, then later liver juice, and even later, liver extract. The lifesaving ingredient in the liver that corrected the deficiency, however, wasn't identified until 1938 by a team led by Merck's Karl Folkers: vitamin B_{12}.

A dozen different ailments, all of them serious, can be helped by adding B_{12} to the diet, from inflammation of the gastrointestinal tract, to immune system disorders like lupus, to celiac disease. As the medical community learned of the lifesaving properties of more and more vitamins, however, they made a monumental though understandable

---

* By convention, thirteen compounds are recognized as essential vitamins in humans. Diseases like scurvy, caused by a deficiency in vitamin C, were recognized for centuries before 1913, when the first vitamin—retinol, a component of vitamin A—was identified (the last, vitamin B_9, was discovered in 1941).

error. Although a deficiency of a particular vitamin was frequently the cause of serious disease—scurvy, from a lack of vitamin C, for example—the reverse isn't true. Taking fewer vitamins than the body needs causes illness; taking more than it needs doesn't promote health. This proved a difficult concept for people, including most doctors, and explains to this day why some of us take five times the recommended daily allowance of vitamin C in the mistaken belief that it will ward off colds. We want to believe that whenever some is good, more must be better. This is also why dozens of biochemists in the 1940s and 1950s dosed domestic animals with massive quantities of vitamins, particularly the "animal protein factor" known as B_{12}.

The search for an inexpensive compound that would make for healthier cows, sheep, pigs, and chickens was well under way by the end of the Second World War, and the researchers working for the newly ambitious pharmaceutical companies were as eager to discover it as they were new weapons in the battle against disease. One such biochemist at Lederle, Thomas H. Jukes, had learned from a review of published research that Merck had discovered that Waksman's world-changing actinomycete, *S. griseus*, produced not just streptomycin, but also B_{12}. He hoped his own search for soil-dwelling bacteria would hit a similar double jackpot, and in December 1948 Jukes received a sample of Aureomycin to test on chickens as a possible cheaper source for the animal protein factor than liver extract, one that was cheap enough for animal feed. The sample was extremely small: enough for Jukes, and his colleague Robert Stokstad, to give supplements to a few laying hens that had been starved to the point that their eggs would normally die within two weeks. The hope was that Aureomycin-produced B_{12}, the protein factor, would allow the chicks to survive.

The results surprised everyone. A dozen young birds had been given Aureomycin, others liver extract, and a control group nothing at all. The chicks given the Aureomycin grew at a rate far faster than those whose hens had been fed a normal diet, which was expected. But they also grew faster than the chicks dosed with liver extract. The reason couldn't be B_{12}. Something in the antibiotic itself was accelerating their growth.* *Streptomyces aureofaciens* had proved to be a gold maker in an entirely different realm.

The demand for Aureomycin as a human antibiotic was so great that Lederle could provide no more of the product to Jukes and his team. In a burst of resourcefulness fully worthy of Norman Heatley at the Dunn School's penurious best, Jukes "dug residues out of the Lederle dump" in order to extract tiny quantities of Aureomycin from the waste products of the fermentation process by which the company manufactured it. He sent the resulting samples to agronomists all over the country to verify its effect, and when a researcher at the University of Florida reported that he had tripled the growth rate of young pigs, even Lederle was convinced. They started marketing Aureomycin to farmers, not as an antibiotic per se, but as a source of B_{12} in order to avoid the annoying regulatory hand of the Food and Drug Administration.

Stokstad and Jukes presented their findings at the annual meeting of the American Chemical Society on April 9, 1950. The following day, the front page of the *New York Times* announced that the new "Wonder Drug . . . had been found to be one of the greatest growth-promoting substances so far to be discovered." They further reported that the use of Aureomycin as a supplement represented "enormous

---

* The mechanism for this is still not very well understood. The initial hypothesis—that the antibiotics reduced the amount of pathogenic bacteria in the gastrointestinal tracts of newborn chickens—hasn't been borne out by subsequent research. A currently popular theory is that GPAs, or growth-promoting antibiotics, work by inhibiting the *normal* intestinal flora that inhabit the GI tracts of domestic animals, and by doing so reduce the "costs" to the animals of digestion. The less energy the chicks used for digesting food, the more they have available for growth. Another currently popular theory is that GPAs (sometimes STAT, for subtherapeutic antibiotic treatment) alter the animal's microbiome: the complex ecosystem of bacteria that live within animals, frequently in numbers that dwarf the number of animal cells with which they coexist, generally not just peaceably, but productively. By killing some percentage of the bacteria in the digestive system, antibiotics, according to this theory, increase the amount of digestible food, and so promote weight gain.

long-range significance for the survival of the human race in a world of dwindling resources and expanding populations." The *Times* even speculated that the drug's "hitherto unsuspected nutritional powers" might aid the growth of malnourished and underweight children.

They wrote truer than they knew. That same year, in one of the more disturbing sidebars to the antibiotic revolution, a Florida physician named Charles Carter started a three-year-long study during which he gave a daily dose of 75 milligrams of Aureomycin to mentally disabled children. As a moment in history when even a wealthy country like the United States worried more about food scarcity than obesity (and hardly at all about the ethics of experimentation on the mentally disabled), Jukes could proudly announce, "The average weight for the supplemented group was 6.5 pounds, while the control group averaged 1.9 pounds in yearly weight gain."

The use of Aureomycin as a method for increasing the growth rate of children did not, luckily, catch on. Not so with domestic animals. Because of the peace treaty that settled the tetracycline patent competition, Lederle wouldn't be alone in pursuing the agricultural market with tetracycline-derived nutritional supplements. In 1950, Pfizer had already put its corporate toe in the water with a compound they named Bi-Con, which combined streptomycin with vitamin B_{12} as a growth additive. If Aureomycin was being profitably sold to farmers, why not Terramycin? In 1952, Herb Luther, an animal nutritionist at Pfizer, started what he called "Project Piglet" in an attempt to find a feed that could be given directly to young pigs, accelerating their growth in order to limit the risk of sows rolling over and crushing their too-small offspring. (Morbid fact: In 1950, more than a third of newborn piglets died each year in this way.) Luther's experimental animals, who were awakened to the stirring sounds of the "William Tell Overture," and put to sleep to "Brahms' Lullaby," were fed on the nursing formula that Pfizer branded as Terralac. Jukes had opened another front in the antibiotic revolution, and here the battlefields wouldn't be pharmacies and hospitals, but feed stores.

It would be more than a little disrespectful to discuss Thomas Jukes without mentioning that he was also a groundbreaking evolutionary biologist, one of the pioneers of what has become a mainstream element of current thinking about molecular evolution—the powerful idea that much evolutionary change is not adaptive, but neutral. This theory, which Jukes first called non-Darwinian evolution, was independently proposed by the Japanese biologist Motoo Kimura in 1968, holds that most evolutionary variability isn't due to natural selection, in which adaptations spread because they improve survival or reproductive productivity, but the random drift of mutations that aren't particularly "superior." In other words, most of the changes that are observed in a species over time are inconsquential; they do nothing to improve survival or reproduction.*

Though Jukes's reputation in the world of biology rests mostly on his work on evolution, he was best known during his lifetime as a journalist who, for more than four decades, fought against pseudoscience and creationism from the bully pulpit of a regular column in *Nature* magazine. In addition, he was a polemicist and public intellectual well remembered today for battling against proposed bans on the insecticide DDT, arguing that the number of lives it saved (by killing malarial mosquitoes) was dramatically greater than any possible ecological risk.

But it was his discovery that antibiotics accelerated the growth of meat-producing animals that has had, by far, the longest tail of consequence. It's not simply, or even mostly, that it led to the increased consumption of animal protein around the world at a dizzying rate. Far more important, it exposed untold quintillions of bacterial pathogens to antibiotics in doses too small to kill them. The result was the cultivation of some of the most robust bacteria the planet has encountered in the last billion years.

The causes of antibiotic resistance are, of course, much greater and more varied than the promiscuous use of drugs like Aureomycin in animal feed. Even before penicillin had been completely isolated or tested on

---

* Such neutral changes are a little like acquiring books for a library that have no immediate impact on the reader's life, but which nonetheless represent a potential future asset.

humans, Ernst Chain and E. P. Abraham had identified an enzyme produced by *Staphylococcus aureus*—penicillinase—that cleaves the chemical bonds holding the beta-lactam ring together, and so degrades the antibiotic action of nearly an entire family of antibiotics.* It wasn't until 1945, though, that the phenomenon was documented: An Australian study tested 159 different strains of *S. aureus*, 128 collected before the advent of penicillin, the other 31 from hospital wards where penicillin was used therapeutically. Only the 31 showed resistance.

But this was just one study, and the first flush of the enthusiasm for the miracle drugs ignored problems of resistance. Articles published during the first years of widespread antibiotic use, roughly 1944 to 1948, seemed to assume that the practice of medicine had largely become a matter of sorting through an array of wonder drugs, and, when in doubt, prescribing them all. Given the uncritical acceptance of the first antibiotics by clinicians—an understandable bit of blinkered thinking, since even when antibiotics were ineffective, they were generally very safe—and the widespread appetite for them by patients, antibiotic resistance was certain to appear, and to appear quickly.

Nonetheless, it was dramatically accelerated by Jukes's discovery. Since their introduction, up to a quarter of all the antibiotics manufactured have been administered to animals. Sometimes those drugs were restricted to animal use; in May 1940, milk cows on display at the "Foodzone" exhibit of the 1939 World's Fair—including the first of Borden's "Elsies"—came down with the mammary gland infection known as mastitis, which was cured by the too-toxic-for-human-use gramicidin. Much more frequently, however, the antibiotics prescribed

* From the 1950s, synthesizing new compounds that were penicillinase resistant led to the discoveries of oxacillin, methicillin, and other drugs with a narrower spectrum of activity than the penicillin G that was first produced in 1945. One of the most important beta-lactam antibiotics, cephalosporin C, was isolated by Abraham himself in 1955, and its structure published in 1961 (and decisively demonstrated by Dorothy Crowfoot Hodgkin's X-ray crystallography). Cephalosporin was not merely effective against bacteria that had mastered the trick of producing beta-lactamase, but was also patentable. Unlike his Dunn School colleagues of a previous generation, Abraham was able to patent his discovery and earned enough from it to endow the Cephalosporin Trust at Oxford's Lincoln College with an initial bequest of £30 million.

were identical to those used by humans. By the 1950s, as much as 10 percent of the milk consumed in the United Kingdom and the United States was contaminated with penicillin, which had been used to treat mastitis in cows who were then milked before the penicillin cleared their systems.

And that was just when antibiotics were used more or less as intended. Because Aureomycin and Terralac were given not only to cure (or even prevent) infection, but to promote growth, the dosages were very small: generally around 200 grams per ton of feed for a two-week period—enough to grow animals to their slaughterhouse weight as quickly as possible, but far below the level needed to stop a bacterial infection. In fact, these exposures more closely replicate the concentrations of antibiotic molecules in nature, such as Alexander Fleming's mold juice or Selman Waksman's soils.

There, they behave very differently. The standard explanation for the phenomenon of bacterial effectiveness—the fact that some microbes produce substances that are highly toxic to other microbes—was historically couched in distinctly martial, not to say anthropomorphic, terms: Bacteria (and fungi like the penicilliums) produce the molecules we call antibiotics to defend themselves in an eternal Hobbesian war of all against all.

Plausible, but for the uncomfortable fact that the molecules that microbes produce aren't usually toxic in the concentrations that occur in nature.

The natural history of antibiotics is not, it turns out, the obvious one, in which unicellular life evolved defense mechanisms during billions of years of natural selection. Antibiotic molecules in nature aren't always, or even usually, weapons. Bacterial populations react to the presence of low concentrations of antibiotic molecules in a variety of ways, and many of them are actually positive. Many antibiotics are what biologists call "hormetic": beneficial in low doses, even promoting the creation of what are known as biofilms, matrices that hold bacterial cells together with a polymer "glue" and make them far more durable in the presence of both the animal immune system, and antibiotics themselves.

This is especially true for the tetracyclines. Low doses of Aureomycin and Terramycin and all the other antibiotics related to them actually increase the virulence of large numbers of pathogens. Subtherapeutic doses of tetracycline help bacteria to form what is known as the type III secretion system, which is one of the key elements of any pathogen's arsenal: a tiny hypodermic needle that Gram-negative bacteria—salmonella, chlamydia, even the organism responsible for bubonic plague—use to inject themselves into animals cells.

Which is one reason why feeding subtherapeutic doses of antibiotics to uncounted millions of pigs, cows, and chickens seems to have been a riskier than optimal strategy.* It might have been designed to select resistant organisms, killing only the "weakest" bacteria—that is, those with the least resistance to the various antibacterial actions—and allowing the stronger to survive. Because so many of the bacteria exposed to low doses of antibiotics had a hormetic reaction to them, the surviving bacteria frequently acquired an even wider arsenal of pathogenic weapons. At the dosages used to promote growth in domestic animals, tetracycline doesn't even slow down an organism like *Pseudomonas aeruginosa,* the bacterium responsible for dozens of opportunistic infections, from pneumonia to septic shock. It strengthens it.

In the early 1950s, however, this was still a problem beyond the time horizon of the pharmaceutical companies, to say nothing of physicians, patients, and regulatory agencies that, if they cared about antibiotics in the food supply at all, were happy to see increased beef and pork production. Antibiotics, particularly broad-spectrum antibiotics, were regarded as an unmixed blessing, the final triumph of medicine over infectious disease: a miracle, one that Americans and Europeans were starting to take for granted.

A society that comes to expect miracles will, sooner or later, have their expectations disappointed.

---

* It's not even a clear advantage in purely economic terms. A 2007 study of chickens raised on subtherapeutic doses of antibiotics showed that the higher cost of the feed outweighed—sorry—any increase in the value of the (somewhat) larger chickens.

EIGHT

"The Little Stranger"

After Selman Waksman and Albert Schatz's discovery, every pharmaceutical company on the planet began obsessively collecting dirt from as many exotic locations as possible in order to improve the odds of finding the next wonder drug in situ. The method was replicated again and again, because it worked.

The organism that produced the crude exudate that would become the world's next great antibiotic was discovered by Abelardo Aguilar, a Filipino physician employed by Eli Lilly, who found a promising soil sample in the country's Iloilo province and, in 1949, sent it to James McGuire at Eli Lilly's Indianapolis headquarters for testing. The sample contained yet another of the ridiculously fecund *Streptomyces*, this time *S. erythreus*: the source for erythromycin, the first of the macrolide antibiotics, compounds that were effective against the same Gram-positive pathogens as penicillin, though via a different mechanism: The macrolides act by inhibiting the way the pathogens make critical proteins, rather than by corroding their cell walls.

Lilly was, compared to upstarts like Pfizer and Merck, very much an old-line American drug company. The firm had been founded in 1876 by Colonel Eli Lilly as "the only House in the West devoted exclusively to the Manufacture and Sale of PHARMACEUTICAL GOODS," which, at the time, largely consisted of botanicals and herbals containing components with memorable names like Bear's Food, Scullcup, and Wormseed. His grandson, also named Eli Lilly, graduated from the Philadelphia College of Pharmacy in 1907 (the same year Paul Ehrlich first described a drug that would attack disease-causing microbes without killing their hosts as a "magic bullet") and joined the company as the de facto director of production shortly thereafter. The younger Lilly was, in some ways, a midwestern version of George Merck: a disciplined and successful twentieth-century industrialist whose enthusiasm for his corporation's larger mission seems to have been as uncomplicated, sincere, and sentimental as the poetry of his fellow Hoosier James Whitcomb Riley. During the Second World War, when the company was furiously producing plasma for American troops overseas, he famously observed that he "didn't think it was the right thing for anybody to make any profit on blood which has been donated."

It has been easy for Lilly's biographers to emphasize his wide though wonky interests. An early enthusiast for the time-motion studies of Frederick Winslow Taylor, Lilly was also, in no particular order, an amateur archaeologist with a special interest in the Native American cultures of his much-loved home state of Indiana; a sophisticated art collector, largely of Chinese paintings and pottery; a compulsive writer of childish rhymes; a devotee of uplifting self-improvement manuals and the music of Stephen Foster; and, for decades, the patron of choice for leaders of now-forgotten academic fads.* He was also, during his lifetime, one of the half dozen most generous American philanthropists. If he did nothing else than initiate a meeting with Frederick Banting, J. J. R. Macleod, and Charles Best at the University of Toronto in 1922, his legacy would be secure. Lilly persuaded them to join in what was then a groundbreaking partnership in developing their discovery—insulin—into a commercial product, which the company launched as Illetin in 1923. For doing well by doing good, it's hard to top; as late as 1975, Eli Lilly still supplied the lifesaving compound to three-quarters of the entire American market.

Insulin was scarcely Lilly's only great innovation prior to the 1950s. Through the 1940s, the company's labs produced the sedative Tuinal and Merthiolate, a widely used antiseptic. Sales rose from

---

* The most prominent were the émigré Russian sociologist Pitirim Sorokin and the evangelical psychologist Ernest Ligon. Though largely forgotten today, during the 1930s Ligon briefly became world famous for his "Character Research Project," a Christian-themed combination of values education and personality measurement.

$13 million in 1932 to $115 million in 1948 (a year in which Eli Lilly thought the profits—21.7 percent—were "unreasonably high"). But while the company was part of the penicillin consortium, and was, for a time, the number one distributor selling Merck's version of streptomycin, they were not one of the major players in the first wave of the antibiotic revolution—not until Abelardo Aguilar's samples arrived in Indianapolis.

Three years later, McGuire applied for a patent on the new drug, which he named erythromycin, a "novel compound having antibiotic properties." It would take decades before the compound could be successfully synthesized; Robert Burns Woodward (who would be credited, posthumously, with solving the problem) wrote in 1956, "Erythromycin, with all our advantages, looks at present quite hopelessly complex," but that did nothing to dissuade Lilly from producing it using the tried-and-true fermentation method. In 1953, Lilly started selling the drug under the name Ilotycin.

Erythromycin was, and would remain, a powerful weapon in the battle against infectious disease. But narrow-spectrum antibiotics like Ilotycin were never going to attract the level of enthusiasm of the new broad-spectrum drugs like the tetracyclines. Broad-spectrum antibiotics accounted for large percentages—as much as half—of the profits of Pfizer, Abbott, Bristol Laboratories, Squibb, and Upjohn. Those five companies, all of whom were selling versions of tetracycline, were splitting about two-thirds of the total market for broad-spectrum drugs.

The other third? In early 1950, a few months before Pfizer's John McKeen joined with Arthur Sackler to transform drug advertising forever, McKeen offered the marketing rights for the oxytetracycline compound to one of his competitors. Their president, however, turned him down, believing that Terramycin was a direct competitor to his own blockbuster broad-spectrum antibiotic. The drug was Chloromycetin, and the company Parke-Davis.

Parke-Davis was then one of America's oldest and largest manufacturers of ethical drugs, compounds that were subject to patent—and, therefore, confusingly, the opposite of "patent medicines"—clearly labeled, and prescribed by physicians. The company's origins date back to 1866, when Hervey Coke Parke, a onetime copper miner and hardware store owner, joined the Detroit drug business of Dr. Samuel Pearce Duffield. Duffield, like New York's Edward R. Squibb, had started a business to serve the needs of the Grand Army of the Republic during the Civil War: distilling alcohol and selling "ether, sweet spirits of nitre [ethyl nitrate, in a highly alcoholic mixture; the spirits were used to treat colds and flu], liquid ammonium [sic], Hoffman's [sic] anodyne [ether and alcohol, used as a painkiller], mercurial ointment, etc." In 1867, a twenty-two-year-old salesman named George Solomon Davis became the firm's third partner; and, when Duffield retired in 1871, the company was incorporated as Parke-Davis and Company. Parke was its first president; Davis its general manager.

Almost immediately, Parke-Davis established a reputation for seeking out medicines in exotic corners of the globe. In 1871 alone, the company financed expeditions to Central and South America, Mexico, the Pacific Northwest, and the Fiji islands. In January 1885, George Davis read Sigmund Freud's infamous article, "Über Coca," in which the young Viennese neurologist (not yet the father of psychoanalysis) wrote:

> The psychic effect of cocaïnum muriaticum in doses of 0.05–0.10g consists of exhilaration and lasting euphoria, which does not differ in any way from the normal euphoria of a healthy person. . . . One is simply normal, and soon finds it difficult to believe that one is under the influence of any drug at all.

Davis immediately dispatched Henry Rusby, a doctor and self-described "botanist and pharmacognosist," to South America to make "a critical study of the different varieties of coca."

Rusby's expedition, "involving four thousand miles of travel by canoe and raft on the Madeira and Amazon rivers, and occupying eleven months of suffering and danger escaping narrowly from death," became part of the company's founding myth, and the beginning of its prosperity. Shortly after his return, the company was using cocaine in dozens of different Parke-Davis products, including coca-leaf cigarettes, wine of coca, and cocaine inhalants. (Davis even hired Freud himself to perform a comparison of Parke-Davis's cocaine products against Merck's.)

Cocaine made the company, but wasn't its last success story. Within twenty years, it had introduced fifty new botanically based drugs to the (still very informal) United States Pharmacopeia. One of them, *Damiana et Phosphorus cum Nux*—damiana is a psychoactive shrub that grows wild in Texas and Mexico; nux is nux vomica, or strychnine—was guaranteed to "revive sexual existence." Others, marketed as "Duffield's Concentrated Medicinal Fluid Extracts," included ingredients like aconite, belladonna, ergot (the cause of St. Anthony's Fire), arsenic, and mercury. All were extremely pure—the firm's motto was *Medicamenta Vera*: "True Medicines"—but are also reminders of the danger in believing "natural" equals "safe." Virtually every page in the catalog of Parke-Davis medications included a compound as hazardous as dynamite, though far less useful.

By the early twentieth century, the company had expanded into areas slightly less dependent on Indiana Jones–like adventuring. In the late 1890s, they were selling their version of Emil Behring's diphtheria antiserum; in 1900, a Parke-Davis chemist, the Japanese-born, Glasgow-educated Jokichi Takamine, isolated adrenaline (also known as epinephrine), which the company marketed as Adrenalin, a drug whose ability to constrict blood vessels made it invaluable to surgeons, especially eye surgeons. The company expanded nationally and internationally, opening offices in Canada, Britain, Australia, India, and, in 1902, opened the country's first full-scale pharmaceutical research laboratory, blocks from their Detroit headquarters. In 1938, Parke-Davis introduced Dilantin, the first reliable treatment for epilepsy; in 1946, Benadryl, the first effective antihistamine, which had been developed by a onetime University of Cincinnati chemist named George Rieveschl, who left academia for a research position at Parke-Davis.

Davis and Parke's successors never lost a taste for treasure hunting, which might explain why they were uninterested in Pfizer's offer. They had a broad-spectrum antibiotic of their own.

The development of Parke-Davis's signature antibiotic began at roughly the same moment in time that the Office of Scientific Research and Development was assembling the participants in the penicillin project. In July 1943, Oliver Kamm, Parke-Davis's director of research, met Paul Burkholder, the Eaton professor of botany at Yale. Six months later, Parke-Davis agreed to fund his research.

It was only a little more than a year before the company's investment in Burkholder—and its long-standing presence in South America—paid off. In April 1945, a month before the surrender of Germany, Derald George Langham,* a plant geneticist simultaneously teaching at the University of Caracas and working as a Parke-Davis consultant, sent Burkholder a crate full of bottles containing compost he had collected from the farm of an émigré Basque farmer named Don Juan Equiram. Hundreds of different soil-dwelling bacteria were isolated from the sample. Most of them were familiar, as was true, too, of the more than seven thousand samples Burkholder received in a single year. Culture A65, however, was different: an entirely new species, a cousin to Waksman's actinomycetes. Burkholder named it *Streptomyces venezuelae* and proceeded to subject it to more or less the same tests that Waksman and Schatz had performed on their soil dwellers: samples of *S. venezuelae* were placed in vertical strips on an agar-containing Petri dish, while colonies of pathogenic bacteria were aligned horizontally. From the warp and weft, it was hoped, a new antibiotic-producing organism would be woven.

Burkholder sent a colony of *S. venezuelae* to John Ehrlich at Parke-Davis.

By the time Ehrlich joined the company in December 1944, he had already collected degrees in phytopathology, mycology, and forest pathology; had worked for the Bartlett Tree Expert Company as an arborist; and served as deputy director of the penicillin program at the University of Minnesota, where he had led the team irradiating variants of the *Penicillium* mold. At Parke-Davis, he had recruited the

---

* Langham is a double footnote to biological history. Theodosius Dobzhansky also used him to find the fruit flies for his groundbreaking genetic research.

company's entire sales force as field researchers, issuing them plastic bags in which they were told to collect soil samples.* Thousands came in, from golf courses, flower gardens, and riverbeds, but until Burkholder's package arrived, none had yielded anything particularly interesting.

Culture A65 was far more than interesting. Quentin Bartz, one of the company's research chemists, isolated the active ingredient using a proprietary technique developed at Parke-Davis that could rapidly reduce thousands of promising molecules to a few dozen. Bartz mixed cultures of A65 and water with fourteen different solvents, each at different levels of acidity. He then removed the water and solvent, filtered what was left (which told him the size of the molecule), and checked whether it adhered to a specific substance (which told him its likely structure). In March 1946, he had a crystalline substance that was effective against not just Gram-positive pathogens, but Gram-negative pathogens as well. It was well tolerated, highly potent against pathogens that were unaffected by either penicillin or streptomycin, and, as an unexpected bonus, could be taken orally, rather than by injection. The chemists at Parke-Davis gave it a nickname: "the Little Stranger."

By February 1947, they had even better news. Chemist Mildred Rebstock had derived the structure of the A65 molecule, whose key component was a ring of the organic compound nitrobenzene. Nitrobenzene had been used for decades as a precursor to the aniline dyes that had been so important to the original sulfanilamides like Prontosil. And nitrobenzene wasn't just familiar; it was simple. Parke-Davis had found a molecule that was far less structurally complex than penicillin, or streptomycin, or erythromycin. This suggested that, unlike its predecessors (or its immediate successor, chlortetracycline), it had the potential to be synthesized rather than grown in

---

* Such was the impact of Waksman's discovery of streptomycin that this became a common tactic for pharmaceutical researchers, occasionally reaching head-scratching levels. In 1951, Bristol-Myers's annual report had a prepaid business reply envelope sewn in, with instructions to shareholders on how to obtain "a teaspoon of soil, slightly moist but not wet" and send it in for testing.

fermentation tanks. If so, it could be produced at considerably less expense and, more important, far greater consistency. By November, Rebstock delivered a completely synthetic and active version of the molecule, which was generically known as chloramphenicol. The company named it Chloromycetin.

Each of the golden age antibiotics, from their introduction to today, is most closely associated from its moment of discovery with a hitherto untreatable disease. As penicillin performed its first miracles on septicemia, and streptomycin was the long-awaited cure for tuber- culosis, Chloromycetin (or chloramphenicol) was greeted enthusiasti- cally mostly because of its activity against insect-borne bacterial diseases, particularly typhus.

Epidemic typhus is a most adept and subtle killer. Victims unlucky enough to encounter a louse carrying a colony of the Gram-negative bacteria known as *Rickettsia prowazekii** typically infect themselves: The lice carry bacteria in their digestive systems and excrete them when they defecate. Humans scratch the lice bites, thus sneaking the pathogen-carrying feces past their skin and into their bloodstreams. The result is the appearance of flu-like symptoms within days: fevers, chills, and aches. A few days later, a rash appears on the victim's torso and rapidly spreads to arms and legs. Then, if the immune system fails to destroy it, the disease progresses to acute meningoencephalitis: an inflammation that simultaneously attacks both the membranes sur- rounding the brain and spinal cord, and the brain itself, causing de- lirium and light sensitivity, eventually leading to coma. If untreated, typhus kills between 10 and 60 percent of those infected.

Typhus has been a scourge of humanity since at least the fifteenth century, and very likely for many centuries before. Epidemics were common throughout early modern Europe, especially in conditions where large numbers of susceptible hosts were placed in the path of lice, such as prisons and among armies on campaign.† During the

---

* Typhus is unrelated to typhoid fever, which is caused by a *Salmonella* bacterium.
† The many informal names for epidemic typhus include "ship fever," "jail fever," and "camp fever."

Thirty Years' War, typhus killed as many as one German in ten. A little less than two centuries later, it killed more soldiers in Napoleon's Grande Armée during the retreat from Moscow than the Russian army. A century after that, a typhus epidemic in the new Soviet Union produced more than twenty million cases, and at least two million fatalities.

The U.S. Army had a long history of concern about the impact of epidemic typhus. The Army Medical Corps dusted more than a million Neapolitan civilians with lice-killing powder enriched with DDT in 1943 out of fear of a typhus outbreak, and the fear didn't vanish at the end of the war. So when the army learned that Parke-Davis had a promising rickettsial antibiotic under development, they were eager to put it through its paces. From late 1946 to early 1947, Dr. Joseph Smadel of the Department of Virus and Rickettsial Diseases at Walter Reed Army Hospital ran A65 through a series of animal experiments, followed by clinical trials. In December 1947, he and two other Walter Reed physicians dosed themselves over a ten-day period with pills of the newly named Chloromycetin in order to discover whether the drug was excreted safely and completely, and more to the point, whether a stable concentration could be maintained in the body. Fortunately for Parke-Davis (and even more so for the physicians themselves), the drug passed both tests with flying colors.

While the Walter Reed doctors were self-testing, Chloromycetin was also getting a field test, one that, given Parke-Davis's history, was taking place in South America. In late November 1947, one of the company's clinical investigators, Dr. Eugene Payne, had arrived in Bolivia, which was then suffering through a typhus epidemic that was killing between 30 and 60 percent of its victims. Payne brought virtually all the chloramphenicol then available in the world (about 200 grams, enough to treat about two dozen patients) and set up a field hospital in Puerto Acosta. Twenty-two patients, all Aymara Indians, were selected for treatment, with another fifty as controls. The results were very nearly miraculous. In hours, patients who had started the day with fevers higher than 105° were sitting up and asking for water. Not a single treated victim died. Out of the fifty members of the control group, only thirty-six survived, a mortality rate of nearly 30 percent.

It was the first of many such field tests. In January 1948, Smadel and a team from Walter Reed recorded similar success in Mexico City. Two months later, they did the same in Kuala Lumpur. Along the way, they discovered that Chloromycetin was effective against the North American rickettsial disease known as Rocky Mountain spotted fever (which can kill more than 20 percent of untreated victims) and the chlamydial disease known variously as parrot fever, or psittacosis. They also found, more or less accidentally*—a patient with typhuslike symptoms turned out to have typhoid instead—that Chloromycetin cured it as well.

The model that had been pioneered by penicillin, and refined for streptomycin and the tetracyclines, was now a well-oiled industrial machine: Microbiologists make a discovery, chemists refine it, and physicians demonstrate its effectiveness in animals and humans. It was time to gear up for industrial production of the new miracle drug. Though the Rebstock experiments had shown how to synthesize the drug (and an improved method had been patented by other Parke-Davis chemists), the drug was still being produced through 1949 both by fermentation and synthesis, the former in a 350,000-square-foot building containing vats originally built to cultivate streptomycin and penicillin.

The process had come a considerable way since those early experiments at the Dunn, and even Pfizer's converted Brooklyn ice factory. A rail line was built directly to the plant and raw materials arrived on a siding, just as if the railroad was delivering steel for an automobile factory. Every week Parke-Davis's workforce unloaded tanker cars full of nutrients like wheat gluten, glycerin, and large quantities of salt; also sulfuric acid, sodium bicarbonate, amyl acetate, and deionized water. The *S. venezuelae* cultures that they were intended to

---

* In the 1940s, the most reliable lab test for typhus, adding a colony of *Proteus* bacteria to the blood of an infected person and examining it microscopically to see if it clumped, found fewer than 40 percent of infections.

feed were produced in separate laboratories, where they were stored until needed in sterilized earth, cultured on demand, suspended in a solution of castile soap, and held in refrigerators.

The feeding process was just as industrial. Nutrient solution was poured into seven 50-gallon steel tanks plated with nickel and chromium as anticorrosives, and then sterilized by heating to 252°. *Streptomyces venezuelae* was then injected, the stew agitated using the same washing-machine technique pioneered at the Northern Lab only a few years before, and held at a controlled temperature of 86° for twenty-four hours. The whole mix was then transferred to 500-gallon tanks, and then to 5,000-gallon tanks—each one seventeen feet high and nearly eight feet in diameter, there to ferment.

Following fermentation, the broth was filtered to remove the no-longer-needed *S. venezuelae* bacteria, thus reducing 5,000 gallons of fermentation broth into 900 gallons of amyl acetate, itself evaporated down to 40 gallons, which was separated and condensed into 2 gallons of solution, from which the antibiotic crystals could—after more than three weeks, and involving hundreds of Parke-Davis chemists, engineers, and technicians—be extracted.

On December 20, 1948, Parke-Davis submitted New Drug Application number 6655 to the Food and Drug Administration, asking that they approve chloramphenicol, and allow the company to bring it to market. On January 12, 1949, the FDA granted the request, authorizing it as "safe and effective when used as indicated." On March 5, 1949, *Collier's* magazine hailed it as "The Greatest Drug Since Penicillin." By 1951, chloramphenicol represented more than 36 percent of the total broad-spectrum business, and Parke-Davis had it all to itself. The Detroit-based company had become the largest pharmaceutical company in the world, with more than $55 million in annual sales from Chloromycetin alone.

This was an enviable position. But also a vulnerable one.

"Blood dyscrasia" is an umbrella term for diseases that attack the complex system by which stem cells in the human bone marrow produce red and white blood cells: erythrocytes, leukocytes, granulocytes, and platelets. Dyscrasias can be specific to one sort—anemia is a deficiency in red blood cells, leukopenia in white—or more than one. Aplastic anemia, a blood dyscrasia first recognized by Paul Ehrlich in 1888, refers to depletion of all of them: of every cellular blood component. The result is not just fatigue from a lack of oxygen distribution to cells, or rapid bruising, but a complete lack of any response to infection. Aplastic anemia effectively shuts down the human immune system.

During the first week of April 1951, Dr. Albe Watkins, a family doctor practicing in the Southern California suburb of Verdugo Hills, submitted a report to the Los Angeles office of the FDA. The subject was the Chloromycetin-caused (he believed) aplastic anemia in his nine-year-old son James, who had received the antibiotic while undergoing kidney surgery, and several times thereafter. On April 7, the LA office kicked it up to Washington, and the agency took notice.

Meanwhile Dr. Watkins, a veteran of the Coast Guard and the U.S. Public Health Service, was making Chloromycetin his life's work: writing to the *Journal of the American Medical Association* and to the president and board of directors of Parke-Davis. His passion was understandable; in May 1952, James Watkins died. Dr. Watkins closed his practice and headed east on a crusade to bring the truth to the FDA and AMA. In every small town and medium-sized city in which he stopped, he called internists, family physicians, and any other MD likely to have prescribed Chloromycetin, carefully documenting their stories.

Albe Watkins was the leading edge of a tidal wave, but he wasn't alone. In January 1952, Dr. Earl Loyd, an internist then working in Jefferson City, Missouri, had published an article in *Antibiotics and Chemotherapy* entitled "Aplastic Anemia Due to Chloramphenicol," which sort of tipped its conclusion. Through the first half of 1952, dozens of clinical reports and even more newspaper articles appeared, almost every one documenting a problem with Chloromycetin. Many of them all but accused Parke-Davis of murdering children.

To say this was received with surprise at Parke-Davis's Detroit headquarters is to badly understate the case. During the three years that Chloromycetin had been licensed for sale, it had been administered to more than four million people, with virtually no side effects.

In the fall of 1952, Albe Watkins made it to Washington, DC, and a meeting with Henry Welch, the director of the FDA's Division of Antibiotics. Dr. Watkins demanded action. He was trying to kick down a door that had already been opened; Welch had already initiated the first FDA-run survey of blood dyscrasias.

The survey's findings were confusing. Detailed information on 410 cases of blood dyscrasia had been collected, but it wasn't clear that chloramphenicol was the cause of any of them. In half the cases—233—the disease had appeared in patients who had never taken the drug. In another 116, additional drugs, sometimes five or more, had been prescribed. Only 61 of the victims took chloramphenicol only, and all of them were, by definition, already sick. The researchers had a numbers problem: Aplastic anemia is a rare enough disease that it barely shows up in populations of less than a few hundred thousand people. As a result, the causes of the disease were very difficult to identify in the 1950s (and remain so today).

Finally, just to further complicate cause and effect, chloramphenicol-caused aplastic anemia, if it existed at all, wasn't dose dependent. This was, to put it mildly, rare; ever since Paracelsus, medicine had recognized that "the dose makes the poison." It not only means that almost everything is toxic in sufficient quantities; it also means that virtually all toxic substances do more damage in higher concentrations. This dose-response relationship was as reliable for most causes of aplastic anemia as for any other ailment. Benzene, for example, which is known to attack the bone marrow, where all blood cells are manufactured, is a reliable dose-related cause of aplastic anemia; when a thousand people breathe air containing benzene in proportions greater than 100 parts per million, aplastic anemia will appear in about ten of them. When the ratio of benzene to air drops below 20 parts per million, though, the incidence of the disease falls off dramatically: only one person in ten thousand will contract it.

Not chloramphenicol, though. A patient who was given five times

more of the drug than another was no more likely to get aplastic anemia. Nor was the drug, like many of the pathogens it was intended to
combat, hormetic—that is, it wasn't beneficial in small doses and only
dangerous in higher ones. The effect was almost frustratingly random. Some people who took chloramphenicol got aplastic anemia.
Most didn't. No one knew why.

Even so, Chester Keefer of Boston University, the chairman of the
Committee on Chemotherapeutics and Other Agents of the National
Research Council during the Second World War (and the man who
had been responsible for penicillin allocation), "felt that the evidence
was reasonably convincing that chloramphenicol caused blood dyscrasias [and that] it was the responsibility of each practicing physician
to familiarize himself with the toxic effects of the drug." In July, after
recruiting the NRC, a branch of the National Academy of Sciences, to
review the findings, FDA Deputy Commissioner George Larrick phoned
Homer Fritsch, an executive vice president at Parke-Davis, to tell him,
"We can't go on certifying that the drug is safe."

Fritsch might have been concerned that the FDA was preparing to
ban Chloromycetin. He needn't have worried, at least not about that.
At the FDA's Ad Hoc Conference on Chloramphenicol, virtually every attendee believed the drug's benefits more than outweighed its
risks. Even Maxwell Finland, who had found the early reports on Aureomycin to be overly enthusiastic, endorsed chloramphenicol's continued use. The Division of Antibiotics recommended new labeling
for the drug, but no restrictions on its distribution. Nor did it recommend any restriction on the ability of doctors to prescribe it as often,
and as promiscuously, as they wished. The sacrosanct principle of
noninterference with physician decisions remained.*

If this sounds like a regulatory agency punting on its responsibility,

---

* To this day, the controversy over the authority of medical professionals to make clinical
decisions has yet to be resolved. The terms that the debate adopts in popular accounts—
should decisions about testing or surgery or drugs be made by "caring physicians" or
"faceless bureaucrats" (whether representing insurance companies or governmental
agencies like the FDA or the Centers for Disease Control and Prevention)—suggest why
the controversy seems unlikely to be resolved soon.

there's a reason. Even with the reforms of 1938, which empowered the FDA to remove a product from sale, the authority to do so had rarely been used. Instead, the agency response, even to a life-threatening or health-threatening risk, was informational: to change the labeling of the drug. In 1953, the FDA issued a warning about the risk of aplastic anemia in the use of chloramphenicol, but offered no guidelines on prescribing.

The result was confusion. A 1954 survey by the American Medical Association, in which 1,448 instances of anemia were collected and analyzed, found "no statistical inferences can be drawn from the data collected." And, in case the message wasn't clear enough, the AMA concluded that restricting chloramphenicol use "would, in fact, be an attempt to regulate the professional activities of physicians."

Except the "professional activities of physicians" were changing so fast as to be unrecognizable. The antibiotic revolution had given medicine a tool kit that—for the first time in history—actually had some impact on infectious disease. Physicians were no longer customizing treatments for their patients. Instead, they had become providers of remedies made by others. Before penicillin, three-quarters of all prescriptions were still compounded by pharmacists using physician-supplied recipes and instructions, with only a quarter ordered directly from a drug catalog. Twelve years later, nine-tenths of all prescribed medicines were for branded products. At the same moment that their ability to treat patients had improved immeasurably, doctors had become completely dependent on others for clinical information about those treatments. Virtually all of the time, the others were pharmaceutical companies.

This isn't to say that the information coming from Parke-Davis was inaccurate, or that clinicians didn't see the drug's effectiveness in their daily practice. Chloromycetin really *was* more widely effective than any other antibiotic on offer: It worked on many more pathogens than penicillin and had far fewer onerous side effects than either streptomycin, which frequently damaged hearing, or tetracycline, which was hard on the digestion.* Chloramphenicol, by comparison, was extremely easy on the patient, with all the benefits and virtually none of the costs of any of its competitors.

Nonetheless, because of the National Research Council report, and the consequent labeling agreement, the company's market position took a serious tumble. Sales of Chloromycetin, which accounted for 40 percent of the company's revenues and nearly three-quarters of its profits, fell off the table. Parke-Davis had spent $3.5 million on a new plant in Holland, Michigan, built exclusively to make the drug; in the aftermath of the report, the plant was idled. The company had to borrow money in order to pay its 1952 tax bill. In September 1953, *Fortune* magazine published an article that described the formerly dignified company as "sprawled on the public curb with an inelegant rip in its striped pants."

Parke-Davis attempted to take the high road, publishing dozens of laudatory studies and estimated the risk of contracting aplastic anemia after taking Chloromycetin at anywhere from 1 in 200,000 to 1 in 400,000.† But the company was fighting with the wrong weapons. In any battle between clinical reports of actual suffering and statistical analyses, the stories were always going to win, especially when the disease in question tends to strike otherwise healthy children and

* Though the phenomenon wasn't well understood until the 1960s, the tetracyclines— because they bind with calcium—can also permanently discolor teeth when given to children.

† The risk of contracting aplastic anemia from chloramphenicol exposure is a constantly moving target. The reason is that it's calculated by dividing the number of cases by the number of people who've been exposed, which is harder to measure than it sounds. Because no one collected data on the number of prescriptions written or filled, the at-risk population can only be inferred by dividing the total amount of chloramphenicol sold in a given year by the average dose, itself an estimate of between 3 and 8 grams for each patient. Thus, the risk was estimated to be 1 in 156,000 in 1958, and 1 in 227,000 in 1959. By 1964, though, a different study calculated the risk at 1 in 60,000. This doesn't show the differential risk of dying from the drug, though. The best guess today is that the risk of dying from aplastic anemia in the absence of chloramphenicol exposure was about 1 in 500,000. For those exposed, the odds dropped to a still rare 1 in 40,000.

adolescents. The papers, articles, and newspaper reports of the day reveal how many of the discoveries of aplastic anemias were specific or anecdotal: Dr. Louis Weinstein of Massachusetts gave a speech before his state medical society in which he revealed he'd heard of—*heard of*—forty cases. The Los Angeles Medical Association reported on two cases, one fatal. Albe Watkins, when he made his famous visit to Henry Welch at the FDA, had collected only twelve documented cases.

Even the objective statistics were problematic. In 1949, the year chloramphenicol was approved for sale, the most reliable number of reported cases of aplastic anemia was 638. Two years later—after millions of patients had received the antibiotic, but before Albe Watkins began his crusade—the number was 671. That increased to 828 in 1952, but most of the 23 percent increase in a single year was almost certainly due to heightened awareness of a disease most physicians—including Albe Watkins—had never encountered before. Even more telling: The increase in blood dyscrasias where chloramphenicol was involved was no greater than where it wasn't. That is, aplastic anemia was on the increase with or without chloramphenicol.

Even though Chloromycetin had such a tenuous cause-and-effect relationship with aplastic anemia, its competitors had no relationship at all. As a result, Parke-Davis was compelled to face the uncomfortable fact that Terramycin and Aureomycin had a similar spectrum of effectiveness, were produced by equally respected companies, and, rightly or wrongly, weren't being mentioned in dozens of newspaper articles and radio stories as a killer of small children.

What really put Parke-Davis in the FDA's crosshairs weren't the gory newspaper headlines, or, for that matter, the NRC study. It was the company's detail men.

The etymology of "detail man" as a synonym for "pharmaceutical sales rep" can't be reliably traced back much earlier than the 1920s. Though both patent medicine manufacturers and ethical pharmaceutical companies like Abbott, Squibb, and Parke-Davis employed salesmen from the 1850s on, their job was unambiguous: to generate direct sales. As such, they weren't always what you might call welcome; in 1902, William Osler, one of the founders of Johns Hopkins and one of

America's most famous and honored physicians, described "the 'drummer' of the drug house" as a "dangerous enemy to the mental virility of the general practitioner."

When Osler wrote that, however, he was describing a model that was already on its way out. Though doctors in the latter half of the nineteenth century frequently dispensed drugs from their offices (and so needed to order them from "drummers"), by the beginning of the twentieth they were far more likely to supply their patients through local pharmacies. Pharmaceutical companies, in response, directed their sales representatives to "detail" them—that is, to provide doctors with detailed information about the company's compounds. By 1929, the term was already in wide circulation; an article in the *Journal of the American Medical Association* observed, "in the past, when medical schools taught much about drugs and little that was scientific about the actions of drugs [that is, doctors never learned why to choose *this* drug rather than *that* one], physicians were inclined to look to the pharmaceutic [sic] 'detail man' for instruction in the use of medicines."

At the time, there were probably only about two thousand detail men in the United States. By the end of the 1930s, there were more of them, but the job itself hadn't changed all that much. In 1940, *Fortune* magazine wrote an article (about Abbott Laboratories) that described the basic bargain behind detailing. In return for forgoing what was, in 1940, still a very lucrative trade in patent medicines, ethical drug companies were allowed a privileged position in their relationships with physicians. They didn't advertise to consumers; their detail men didn't take orders. They weren't anything as low-rent as "salesmen."

Or so they presented themselves to the physicians whose prescription pads were the critical first stop on the way to an actual sale. In the same year as the *Fortune* article, Tom Jones (a detail man for an unnamed company) wrote a book of instructions for his colleagues, in which he cheerfully admitted, "Detailing is, in reality, sales promotion, and every detail man should keep that fact constantly in mind."

With the antibiotic revolution of the 1940s, the process of detailing, and the importance of the detail man, changed dramatically. A 1949 manual for detail men (in which they were described as "Professional Service Pharmacists") argued, "The well-informed 'detail-man' is one of the most influential and highly respected individuals in the public-health professions. His niche is an extremely important one in the dissemination of scientific information to the medical, pharmaceutical, and allied professions. . . . He serves humanity well."

He certainly provided a service to doctors. In 1950, about 230,000 physicians were practicing in the United States, and the overwhelming majority had left medical school well before the first antibiotics appeared. This didn't mean they hadn't completed a rigorous course of study. The 1910 Flexner Report—a Carnegie Foundation–funded, American Medical Association–endorsed review of the 155 medical schools then operating in the United States—had turned medical education into a highly professional endeavor.* But while doctors, ever since Flexner, had been taught a huge number of scientific facts (one of the less than revolutionary recommendations of the report was that medical education be grounded in science), few had really been taught how those facts had been discovered. Doctors, then and now, aren't required to perform scientific research or evaluate scientific results.

Before the first antibiotics appeared, this wasn't an insuperable problem, at least as it affected treating disease. Since so few drugs worked, the successful practice of medicine didn't depend on picking the best ones. After penicillin, streptomycin, and chloramphenicol, though, the information gap separating pharmaceutical companies from clinicians became not only huge, but hugely significant. Detail men were supplied with the most up-to-date information on the effectiveness of their products—not only company research, but also

---

* Most of Flexner's recommendations are jaw-droppingly obvious in retrospect. One required physicians to have completed at least two years of college (only sixteen schools required this at the time of the study). Another proposed that they receive clinical training in addition to lectures. A third recommended that medical schools, following the European model, should affiliate with universities, rather than operate as stand-alone, for-profit vocational schools. The result, once the AMA's Council on Medical Education persuaded each of the state licensing boards of the merit of the report, was to cut the number of accredited medical schools in the United States by half.

reprints of journal articles, testimonials from respected institutions and practitioners, and even FDA reports. Doctors, except for those in academic or research settings, weren't. In 1955, William Bean, the head of internal medicine at the University of Iowa College of Medicine, wrote, "A generation of physicians whose orientation fell between therapeutic nihilism and the uncritical employment of ever-changing placebos was ill prepared to handle a baffling array of really powerful compounds [such as the] advent of sulfa drugs, [and] the emergence of effective antibiotics. . . ."

The detail man was there to remove any possibility of confusion. And if, along the way, he could improve his employer's bottom line, all the better. As the same 1949 manual put it, "The Professional Service Pharmacist's job is one of scientific selling in every sense of the word. . . . He must be a *salesman* first, last, and always."

In general, doctors in clinical practice thought the bargain a fair one. Detail men were typically welcomed as pleasant and well-educated information providers, who, incidentally, also provided free pens, lunches, and office calendars in quantity.* Parke-Davis, in particular, hired only certified pharmacists for their own detailing force, and it was said that a visit from one of them was the equivalent of a seminar in pharmacology.

In 1953, when the Chloromycetin story blew up, Harry Loynd was fifty-five years old, and had spent most of his adult life selling drugs, from his first part-time job at a local drugstore to a position as pharmacist and store manager in the Owl Drug Company chain. He joined Parke-Davis as a detail man in 1931, eventually rising to replace Alexander Lescohier as the company's president in 1951. He was aggressive, disciplined, autocratic, impatient with mistakes, and possessed of enormous energy.

However, unlike his predecessor or most of his fellow industry leaders, Loynd had little use for the medical profession. At one sales

---

* The level of what an objective observer might call unethical gift giving would, with the number of branded drugs, increase to stratospheric levels soon enough. See Chapter Nine.

meeting, surrounded by his beloved detail men, he told them "If we put horse manure in a capsule, we could sell it to 95 percent of these doctors." And when he said, "sell" he didn't mean "advertise." Ads in magazines like *JAMA* were fine, in their place; they were an efficient way of reaching large numbers of physicians and other decision makers. But advertising wasn't able to build relationships, or counter objections, or identify needs. For that, there was nothing like old-fashioned, face-to-face selling. Parke-Davis wouldn't use the clever folks at the William Douglas McAdams ad agency. Loynd was a salesman through and through and believed that Parke-Davis's sales force wasn't just a source of its credibility to doctors, but its biggest competitive advantage.

Even before the FDA announced its labeling decision, Loynd was spinning it as a victory, issuing a press release that said—accurately, if not exhaustively so—"Chloromycetin has been officially cleared by the FDA and the National Research Council with *no restrictions* [italics in original] on the number or the range of diseases for which Chloromycetin can be administered. . . ." Doctors all over the country received a letter using similar language, plus the implication that other drugs were just as complicit in cases of aplastic anemia as Chloromycetin. Most important: The Parke-Davis sales force was informed, apparently with a straight face, that that National Research Council report was "undoubtedly the highest compliment ever tendered the medical staff of our Company." Parke-Davis would use its detail men to retake the ground lost by its most important product.

Loynd's instinct for solving every problem with more and better sales calls was itself a problem. When management informs its sales representatives that they are the most important people in the entire company—Loynd regularly told his detail men that the only jobs worth having at Parke-Davis were theirs . . . and his—they tend to take it to heart. Though the company did all the expected things to get its detail men to tell doctors about the risks of Chloromycetin, even requiring every sales call to end with the drug's brochure open to the page that advised physicians that the drug could cause aplastic anemia, there was only so much that could be done to control every word that came from every sales rep's mouth. Detail men were salesmen "first, last, and always," and more than 40 percent of their income came from a single product. Expecting them to emphasize risks over benefits was almost certainly asking too much.

The FDA, which was asking precisely that, was infuriated. The agency's primary tool for protecting public safety was controlling the way information was communicated to doctors and pharmacists. They could review advertising and insist on specific kinds of labeling. They could do little, though, about what the industry—and especially Parke-Davis—regarded as its most effective communication channel: detail men. It's difficult to tell whether the FDA singled out Parke-Davis for special oversight. In one telling example, a San Francisco physician accused two of the company's detail men of promoting Chloromycetin using deceptive statements at a meeting set up at the FDA's regional office. But there's no doubt that Parke-Davis believed it to be true.

For the next five years, the company walked the narrow line between promoting its most important product and being the primary source of information about its dangers. By most measures, they did it extraordinarily well. Sales recovered—production of the drug peaked at more than 84,000 pounds in 1956—even as it had to survive a second public relations nightmare. In 1959, doctors in half a dozen hospitals started noting an alarming rise in neonatal deaths among infants who had been given a prophylactic regimen of chloramphenicol because they were perceived to be at higher than normal risk of infection, usually because they were born premature. Those given chloramphenicol either alone, or in combination with other antibiotics such as penicillin or streptomycin, were dying at a rate five times higher than expected. The cause was the inability of some infants to metabolize and excrete the antibiotic. It's still not well understood why some infants had this inability, but in a perverse combination, the infants receiving chloramphenicol not only were the ones at most risk, but once they developed symptoms of what has come to be known as "gray baby syndrome"—low blood pressure, cyanosis, ashy skin color—they were given larger and larger doses of the drug. Gray babies frequently showed chloramphenicol blood levels five times higher than the acceptable therapeutic dose.*

Gray baby syndrome was bad enough. Aplastic anemia was worse, and it was the risk of that disease that returned Chloromycetin to the news in the early 1960s. The new aplastic anemia scare was fueled in large part by the efforts of a Southern California newspaper publisher named Edgar Elfstrom, whose daughter had died of the disease after being treated—overtreated, really; a series of doctors prescribed more than twenty doses of Chloromycetin, one of them intravenously—for a sore throat. Elfstrom, like Albe Watkins before him, made opposition to chloramphenicol a crusade, and he had a much bigger trumpet with which to rally his troops. Watkins had been a well-respected but little-known doctor. Elfstrom was a media-savvy writer, editor, and newspaper publisher. He sued Parke-Davis and his daughter's physicians; he wrote dozens of open letters to FDA officials, to members of Congress, to Attorney General Robert Kennedy, and to Abraham Ribicoff, the secretary of the Department of Health, Education, and Welfare. He even met with the president. As someone with easy access to the world of print journalism—Elfstrom wasn't just a publisher himself, but a veteran of both UPI and the Scripps Howard chain of newspapers, with hundreds of friends at publications all over the country—he was able to give the issue enormous prominence. For months, stories appeared in both Elfstrom's paper and those of his longtime colleagues, including a major series in the *Los Angeles Times*. They make heartbreaking reading even today: A teenager who died after six months of chloramphenicol treatment for acne. An

---

* One reason that infants were at particular risk for adverse reactions to a medicine like chloramphenicol is that clinical trials hardly ever include children. Except for cancer research (and rarely even then), parents tend to be unwilling to enroll their children in double-blind randomized studies in which they might be part of a control group. Even today, pediatricians regularly prescribe medications for off-label use, precisely because so few drugs for children can qualify for clinical trials. See Chapter Nine for more on phased clinical trials.

eight-year-old who contracted aplastic anemia after treatment for an ear infection. Four-year-olds. Five-year-olds. A seventeen-year-old with asthma. The stories have a chilling consistency to them: a minor ailment, treatment with a drug thought harmless, followed by subcutaneous bleeding—visible and painful bruising—skin lesions, hemorrhages, hospitalization, a brief respite brought about by transfusions, followed by an agonizing death.

The tragic conclusion to each of these stories is one reason that the chloramphenicol episode is largely remembered today as either a fable of lost innocence—the realization that the miracle of antibacterial therapy came at a profound cost—or as a morality tale of greedy pharmaceutical companies, negligent physicians, and impotent regulators. The real lessons are subtler, and more important.

The first takeaway isn't, despite aplastic anemia and gray babies, that antibiotics were unsafe; it's that after sulfa, penicillin, streptomycin, and the broad-spectrum antibiotics, it wasn't clear what "unsafe" even meant.

For any individual patient, antibiotics were—and are—so safe that a busy physician could prescribe them every day for a decade without ever encountering a reaction worse than a skin rash. It's worth recalling that, only fifteen years before the aplastic anemia scare, the arsenal for treating disease had consisted almost entirely of a list of compounds that were simultaneously ineffective and dangerous. The drugs available at the turn of the twentieth century frequently featured toxic concentrations of belladonna, ergot, a frightening array of opiates, and cocaine. Strychnine, the active ingredient in Parke-Davis's *Damiana et Phosphorus cum Nux*, is such a powerful stimulant that Thomas Hicks won the 1904 Olympic marathon while taking doses of strychnine and egg whites *during the race* (and nearly died as a result). The revolutionary discoveries of Paul Ehrlich and others replaced these old-fashioned ways of poisoning patients with scarcely less dangerous mixtures based on mercury and arsenic. No doctor wanted to return to the days before the antibiotic revolution.

But what was almost certainly safe for a single patient, or even all the patients in a single clinical practice, was just as certainly dangerous to *someone*. If a thousand patients annually were treated with a particular compound that had a 1 in 10,000 chance of killing them, no one was likely to notice the danger for a good long while. Certainly not most physicians. Eight years after Parke-Davis started affixing the first FDA-required warning labels to Chloromycetin, and even after the first accounts of gray baby syndrome, the Council on Drugs of the AMA found that physicians continued to prescribe it for "such conditions as . . . the common cold, bronchial infections, asthma, sore throat, tonsillitis, miscellaneous urinary tract infections . . . gout, eczema, malaise, and iron deficiency anemia." The FDA had insisted on labeling Parke-Davis's flagship product with a warning that advised physicians to use the drug only when utterly necessary, and that hadn't even worked.

Most clinicians simply weren't suited by temperament or training to think about effects that appear only when surveying large populations. They treat individuals, one at a time. The Hippocratic Oath, in both its ancient and modern versions, enjoins physicians to care for patients as individuals, and not for the benefit of society at large. Expecting doctors to think about risk the same way as actuaries was doomed to failure, even as the first antibiotics changed the denominator of the equation—the size of the exposed population—dramatically. Tens of millions of infections were treated with penicillin in 1948 alone; four million people took Chloromycetin, almost all of them safely, from 1948 to 1950.

But if doctors couldn't be expected to make rational decisions about risk, then who? If the chloramphenicol story revealed anything, it was just how poorly society at large was at the same task. As a case in point, while no more than 1 in 40,000 chloramphenicol-taking patients could be expected to contract aplastic anemia, a comparable percentage of patients who took penicillin—1 in 50,000—die from anaphylaxis due to an allergic reaction; and, in 1953, a *lot* more penicillin prescriptions were being written, every one of them without the skull-and-crossbones warning that the FDA had required on Parke-Davis's flagship product.

Chloramphenicol also demonstrated why pharmaceutical companies were severely compromised in judging the safety of their products. As with physicians, this wasn't a moral failing, but an intrinsic aspect of the system: a feature, not a bug. The enormous advances of the antibiotic revolution were a direct consequence of investment by pharmaceutical companies in producing them. The same institutions that had declined to invest hundreds of pounds in the Dunn School's penicillin research were, less than a decade later, spending millions on their own. That, in turn, demanded even greater resources—collecting more and more samples of soil-dwelling bacteria; testing newer and newer methods of chemical synthesis; building larger and larger factories—in improving on, and so replacing, them.

This, as much as anything else, is the second lesson of chloramphenicol. Producing the first version of a miracle drug doesn't have to be an expensive proposition. But the second and third inevitably will be, since they have to be more miraculous than the ones already available. This basic fact guarantees that virtually every medical advance is at risk of rapidly diminishing returns. The first great innovations—the sulfanilamides, penicillin—offer far greater relative benefits than the ones that follow. But the institutions that develop them, whether university laboratories or pharmaceutical companies, don't spend less on the incremental improvements. Precisely because demonstrating an incremental improvement is so difficult, they spend more. The process of drug innovation demands large and risky investments of both money and time, and the organizations that make them have a powerful incentive to calculate risks and benefits in the way that maximizes the drug's use. Despite the public-spiritedness of George Merck or Eli Lilly, drug companies—and, for that matter, academic researchers—were always going to be enthusiasts, not critics, about innovative drugs. It's hard to see how the antibiotic revolution could have occurred otherwise.

This left the job of evaluating antibiotics to institutions that, in theory at least, should have been able to adopt the widest and most disinterested perspective on the value of any new therapy. This was why the Food, Drug, and Cosmetic Act of 1938 empowered the FDA to oversee drug safety—which sounds clear, but really isn't. The decades since the Elixir Sulfanilamide disaster had demonstrated that any drug powerful enough to be useful was, for some patients, also unsafe. Few people, even at the FDA, really understood how to compare risks and benefits in a way that the public could understand.

The third lesson of the chloramphenicol episode should have been that risks and benefits in drug use aren't measured solely by the probability of a bad outcome, or even its magnitude. They can only be established by comparing the risk of *using* a compound against the risk of *not using* it. For this reason, the association of chloramphenicol with blood dyscrasias, while tragic and notorious, was actually beside the point. Chloramphenicol, like penicillin, streptomycin, erythromycin, and the tetracyclines, was an almost unimaginably valuable medicine when used appropriately. The very different incentives of pharmaceutical companies and physicians—the first to maximize the revenue from their investments; the other to choose the most powerful treatments for their patients—practically guaranteed a high level of inappropriate use. Chloramphenicol was critical for treating typhus; not so much for strep throat.*

What the story of chloramphenicol's rise and ultimate decline (at least as an antibiotic prescribed millions of times annually) revealed was that safety couldn't be measured in a vacuum. Evaluating the danger of any new therapy demanded context—its efficacy, as balanced against its risks. The only candidate to do this was the FDA, but the Food, Drug, and Cosmetic Act only empowered the agency to measure safety, not effectiveness.

That was about to change.

---

* Chloramphenicol remains one of the drugs deemed essential by the World Health Organization and is still widely used in many parts of the world for everything from scrub typhus to eye infections. While the drug is still effective, it was supplanted by new and improved antibiotics in industrialized countries—though it did hang on for a fairly long time, and for a perverse reason: Because all the adverse publicity made it less frequently prescribed, it was far slower to develop antibiotic resistance than the beta-lactam or tetracycline antibiotics.

NINE

"Disturbing Proportions"

The inaugural issue of the *Saturday Review of Literature* arrived on newsstands in August 1924, and for the next eighteen years the magazine was edited by Henry Seidel Canby, a professor at Yale University. Canby assembled an impressive group of literary critics to produce the weekly magazine, including the essayist and novelist Christopher Morley, and Mark Twain's biographer Bernard DeVoto. But the *Saturday Review* is best remembered today as the brainchild of Norman Cousins, who took over as editor in 1942. The Cousins era, which lasted until his departure in 1971, represented the magazine's high-water mark in circulation and influence, when it was the voice most attended to by America's midcentury, middlebrow households.

Cousins regularly urged his staff to remember, "There is a need for writers who can restore to writing its powerful tradition of leadership in crisis," and to that end, he and the magazine tirelessly advocated for the entire catalog of largely unreachable liberal objectives, from world government to nuclear disarmament. For actual impact on current events, though, the *Saturday Review* never exhibited a more significant bit of leadership-in-crisis writing than in a series that began in its January 3, 1959, issue.

The cover featured a photograph of Arthur Schlesinger, Jr., whose latest book, *The Coming of the New Deal*, was reviewed. The lead article, written by the magazine's science editor, John Lear, was "Taking the Miracle Out of Miracle Drugs." Its first line read, "Prescription of antibiotics without a specific cause for such treatment has reached disturbing proportions."

The causes of that disturbance, as enumerated by one of Lear's interviewees, Dr. Henry Kempe, head of pediatrics at the University of Colorado Medical School, were several. First was that reflexive prescription of antibiotics often hid the real disease from diagnosticians; during the ten days required for most antibiotic treatments, the patient would frequently grow worse, as the true cause of disease went unaddressed. Second, despite the generally well-tolerated character of most antibiotics, when millions of people are given them every day, thousands will exhibit symptoms of antibiotic poisoning—everything from vomiting to skin rashes. Third, antibiotics, as antagonists to all sorts of bacteria, often caused gastrointestinal upset by killing the "good" bacteria residing in the digestive tract.

But the biggest problem with the miracle drugs was that prescribing them for everything from head colds to migraines was breeding antibiotic resistance. Lear, writing that antibiotic-resistant strains of pathogens had been "known to medicine for almost five years," actually understated the case significantly. Alexander Fleming, in his 1945 Nobel Prize Lecture, had warned, "It is not difficult to make microbes resistant to penicillin in the laboratory by exposing them to concentrations not sufficient to kill them," and it wasn't exactly a newsworthy observation even then. Between 1954 and 1958, as Lear documented, hospitals in the United States had experienced five hundred outbreaks of diseases caused by antibiotic-resistant pathogens—outbreaks that spread fast enough that they met the formal definition of local epidemics: disease episodes in which the daily number of new infections exceeds the number of cases resolved (though, being local, such epidemics were typically not the stuff of newspaper headlines).

The real point of interest, to Lear, was why the epidemics were appearing at all. Since physicians knew (or should have known) that antibiotics were next to useless against viruses, why did they persist in prescribing them for viral diseases? Although it's possible that some physicians knew that most antibiotic activity occurred by disrupting the bacterial cell walls—walls that viruses, which are essentially free-floating bits of DNA, neither have nor need—did they fail to realize that this made viral disease essentially invulnerable to antibiotic treatment? Though doctors had been overshooting the mark on treating patients for decades—Oliver Wendell Holmes, in the same speech in which he recommended consigning the whole materia medica to the seafloor, recognized that "part of the blame of over-medication must, I fear, rest with the profession, for yielding to the tendency to self-delusion which seems inseparable from the practice of the art of healing"—Lear thought the answer was more obvious, and scandalous: advertising. "Established ethical drug companies, traditionally cautious in advancing claims for their medicines, are being jostled and jolted competitively in antibiotic sales by the Madison Avenue 'hard sell' of bulk chemical makers. . . ."

Lear wasn't lacking for specifics. In a section of the article entitled "The Case of the Invisible Physicians," he described a brochure produced by Pfizer, intended for doctors, promoting "the antibiotic formulation with the greatest potential value and the least probable risk . . . Sigmamycin . . . the antibiotic therapy of choice." The promotional piece featured photos of eight business cards, each with the name of a physician. One was from Massachusetts, another from Oregon. The others—including a dermatologist, a urologist, and a pediatrician, in case the message about Sigmamycin's wide range of effectiveness wasn't getting through—practiced in Florida, Arizona, California, Illinois, Pennsylvania, and New York, where each one was an enthusiastic endorser of the drug.

Since the business cards included addresses and phone numbers, Lear tried to contact them. What he found were nonworking phone numbers and addresses from which his letters had been returned, either marked address, or addressee, unknown. The doctors and their testimonials were purest fiction, the "Madison Avenue 'hard sell'" invention of a creative copywriter in the William Douglas McAdams agency.

Lear's piece generated a huge number of letters to the *Saturday Review*, both admiring—"the most factual and intelligent article written on the subject to date"—and critical: "I don't know a single physician who pays the slightest attention to drug ads." Pfizer's president, John McKeen, visited the magazine's editorial offices, where he admitted that the brochure could have been misleading, and that the company had initiated procedures to prevent a recurrence. But the first article had been only a ranging shot. Lear's second would be dead on target. "The Certification of Antibiotics" was the cover story of *Saturday Review's* February 7 issue (in what was surely a coincidence, it appeared just underneath an article by Adlai Stevenson entitled "Politics and Morality"). Its subject was the same senior official in the Food and Drug Administration who had been visited by Albe Watkins in 1952, a bacteriologist named Henry Welch.

Welch had joined the FDA in 1938, as part of the agency's expansion after the passage of the Food, Drug, and Cosmetic Act, and had risen through the ranks fairly quickly. In 1943, he was named to direct the Division of Penicillin Control and Immunology, and in 1951 was appointed director of the Division of Antibiotics, where he was responsible for the approval of new drugs.

At almost precisely the same time Welch took on his new responsibilities at the FDA, he was introduced to a Spanish psychiatrist named Félix Martí-Ibáñez. Martí-Ibáñez had been his country's undersecretary of health and social service until 1939, when the nationalist victory in the Spanish civil war forced him to find other employment. He spent the 1940s working in the United States for a number of pharmaceutical companies, including Hoffmann-La Roche (where he served as medical advisor for international sales), Winthrop, and Squibb. The émigré psychiatrist therefore had a ringside seat for the antibiotic revolution, which he correctly viewed as a huge opportunity: not for creating research, but retailing it. The discovery of antibiotic therapy wasn't just transforming the practice of medicine, but also the practice of medical communication. An unprecedented explosion in useful therapeutic knowledge—it's worth remembering that penicillin, streptomycin, the various versions of tetracycline, chloramphenicol, and erythromycin had all been introduced between 1941 and 1948—was occurring at a rate that made existing channels far too slow. An organization that could promise rapid diffusion to the largest number of clinicians in the shortest amount of time would be supplying something of enormous value.

From the late 1950s on, most academic physicians regarded Martí-Ibáñez as a bit of a huckster, and his reputation hasn't improved very much since. One thing he can't be accused of, however, is hypocrisy. The Spanish psychiatrist made it clear in speeches and articles that the ideal source for financing the organization and diffusion of therapeutic knowledge was the pharmaceutical firms themselves. "Who better than the pharmaceutical industry," he wrote, "could organize, coordinate, and integrate on an international scale the vast and increasing knowledge on antibiotics?"

Martí-Ibáñez had a business plan. In 1951, he put it into action. He and Welch joined forces to found a new journal, entitled *Antibiotics and Chemotherapy*, with an editorial board that included a who's who of antibiotic research, including Florey, Waksman, and Alexander Fleming. Martí-Ibáñez, as president of MD Publications, would run the business side; Welch would be the editor.

Antibiotics and Chemotherapy was an immediate success with medical researchers, but its content was virtually all bench science, which limited its appeal. To serve the much larger audience interested in the clinical application of the new drugs, in 1955 Welch and Martí-Ibáñez launched another journal, which they named *Antibiotic Medicine*, changing the title a year later to *Antibiotic Medicine and Clinical Therapy*. The new journal was circulated free to physicians and other health professionals, as Martí-Ibáñez and Welch reasoned that delivering to that particular audience would make the publication an attractive place for pharmaceutical company advertising. It seems not to have occurred to anyone at the FDA that allowing their antibiotic division's director to work for a for-profit journal supported entirely by the same pharmaceutical companies whose applications he was responsible for endorsing or rejecting was the very definition of a conflict of interest.

For John Lear, alarm bells sounded. When he interviewed Welch for his February 1959 article and asked him to confirm or deny the rumor that he derived substantial income from the journals, Welch replied, "Where my income comes from is my own business [but] I have no financial interest in MD Publications. . . . My only connection is as editor, for which I receive an honorarium."

It would be some months before the size of that honorarium was fully understood. *Antibiotic Medicine and Clinical Therapy* paid Henry Welch 7.5 percent of all advertising revenue, and 50 percent of all sales of article reprints. In the four years between the journal's introduction and John Lear's exposé in *Saturday Review,* the two "honoraria" had paid Henry Welch nearly $250,000, about $2.24 million today. Welch, it was later learned, had told a number of colleagues that his FDA salary—$17,500 a year—was barely enough to pay his income tax. They thought he was kidding.

There is little evidence that Welch was, in the classic sense, corrupted by this. But whether or not he agreed to approve antibiotics produced by journal advertisers, pharmaceutical companies adopted a grateful, and generous, attitude toward MD Publications generally, and Welch personally. Parke-Davis, to mention only a single example, wrote a check for $100,000 for prepaid advertising in *Antibiotic Medicine and Clinical Therapy* during its first year of publication. When the journal folded in 1961, it still had $38,000 of Parke-Davis's money, which the company graciously agreed to write off as a goodwill gesture. Half of the money still in the cash register, $19,000, was paid directly to Henry Welch.

Between 1955 and 1960, Pfizer paid $171,000 for reprints, earning Henry Welch more than $85,000. Even more incriminating: Lear discovered letters to Pfizer's director of advertising in which Welch and Martí-Ibáñez pleaded with him to continue supporting *Antibiotics and Chemotherapy.* The money quote included the following: "The February issue of this publication will include an editorial by Dr. Welch reappraising the use of Nystatin [an antifungal drug derived from yet another of the fecund *Streptomyces,* which Bristol-Myers Squibb was promoting heavily] in conjunction with broad-spectrum antibiotics. *This paper will furnish your people with excellent ammunition with which to counteract the exaggerated claims made for*

Nystatin" [emphasis added]. The quid pro quo—*Antibiotics and Chemotherapy* was about to criticize Nystatin in print; Nystatin's competitor could pay to share that seemingly disinterested criticism with its sales force, and through them with America's physicians—was implied, but clear.

It was an era that was no stranger to outrage over disgraceful behavior, as both the "payola scheme," in which record companies paid disc jockeys for airplay, and the TV quiz show scandals were huge stories in 1959. Henry Welch was, to most people, small potatoes. But that didn't mean no one noticed. As a direct result of John Lear's articles, Congressman Emanuel Celler of New York insisted that Welch should be fired, and soon enough, Dr. Arthur Flemming, the secretary of the Department of Health, Education, and Welfare, demanded his resignation. Welch, who had already applied for disability retirement (and who had a very comfortable retirement, funded by his publishing empire, to look forward to), complied.

To this day, Henry Welch, even more than Félix Martí-Ibáñez, remains a poster boy for the dangers of corruption by pharmaceutical companies, so much so that Web sites with a conspiratorial tinge continue to invoke him as the original "shill for big pharma." They have a point, but they miss the far more significant aspect of Welch's importance to pharmaceutical development in the late 1950s.

Welch and Martí-Ibáñez sincerely believed in the broadest possible use of antibiotics, often in combination not just with other antibiotics, but with vitamins—in a typical example, Pfizer created a compound consisting of Terramycin and "stress formula" vitamins—to both treat and prevent infectious diseases. More to the point, Pfizer had researched, developed, and manufactured Sigmamycin, the combination of 167 milligrams of tetracycline and 83 milligrams of oleandomycin (a close relative of erythromycin) that had been endorsed by the eight nonexistent doctors John Lear had exposed in his January article.

By the mid-1950s, all of the most popular antibiotic therapies were "fixed-dose combination drugs"—mixed cocktails of erythromycin and penicillin, for example. The idea behind them was, on the surface, plausible enough, a long-standing belief in the synergistic combination of two therapies. As far back as 1913, Paul Ehrlich himself had recommended "a simultaneous and varied attack . . . directed at the parasites, in accordance with the military maxim, march in detachments, fight as a unit." Forty-five years later, pharmaceutical companies had taken Ehrlich's "if some is good, more must be better" recommendation to heart, and were producing no fewer than sixty-one fixed-dose combination antibiotics. Four of them contained five antibiotics apiece, eight had four, twenty used three, and "only" twenty-nine were humble enough to combine a mere two. When the drugs in combination were truly synergistic, this therapy worked fine. Sometimes, as with PAS and streptomycin, they were. And sometimes they weren't. Two or more antibiotics could be antagonistic, as was already known to be the case with penicillin and Aureomycin.

Synergistic or not, fixed-dose combinations remained attractive to pharmaceutical companies. Working in a marketplace in which every competitor could sell generic versions of tetracycline or penicillin, even the most innovative firms saw the advantage in differentiating their branded products one from the other. And if they hadn't yet figured it out on their own, Félix Martí-Ibáñez had been eager to educate them. In 1956, he wrote:

> [I]t is particularly important to seek specialties which combine antibiotics with other drugs. Such combinations, if they can be justified medically, are a defense against the current price trend in penicillin and streptomycin. The broad spectrum antibiotics [i.e., the tetracyclines and chloramphenicol] may eventually suffer the same fate. This is the time, therefore, to seek products combining such drugs as Terramycin and Aureomycin with other useful therapeutic agents. . . .

The recipient of the letter wasn't a clinician or researcher. It was Martí-Ibáñez's good friend and fellow psychiatrist, the advertising mastermind at William Douglas McAdams, Arthur Sackler.

The challenge with fixed-dose combination antibiotics, however, was that while it was easy enough for an advertising copywriter to invent a catchy name and package for them, it was very hard indeed to convince physicians and hospitals that a particular fixed-dose combination wasn't just *different*, but *better*. The rigor of well-designed randomized clinical trials made measuring the relative merits of each component of a two-part fixed-dose combination extremely uncertain. With three or more, it was virtually impossible. Further, unless the pathogen involved—and not just the bacterial species, but its biovar, or strain—was identified precisely, the fixed-dose combinations were as likely to harm as heal.

Fixed-dose combination therapies were also a challenge to evaluate effectively using the double-blind randomized clinical trials pioneered by Bradford Hill. In what seems an almost willful return to the preantibiotic era, fixed-dose combinations could only be trumpeted using case studies and, especially, testimonials.

Enter Henry Welch and Félix Martí-Ibáñez. In journal editorials and in speeches at the annual symposia on antibiotics that they hosted throughout the 1950s, they announced that the world had entered a "third era of antibiotic therapy" . . . (the first had been the narrow-spectrum antibiotics like penicillin; the second the broad-spectrum drugs like tetracycline). And if the only way to usher in the "third era"—neither of the two medical publishing innovators felt it necessary to note that the phrase had been provided by one of Arthur Sackler's copywriters, who had coined it for the launch of Sigmamycin—was by replacing randomized clinical trials with personal experiences, then so much the worse for RCTs.

The battle lines had been drawn at the 1956 antibiotics symposium. On one side, Welch and Martí-Ibáñez argued, "The final verdict on the value of a new drug or a new therapy usually comes from one dependable source: the whole body of practicing physicians whose daily clinical experiences extend over many patients treated in actual conditions of practice over considerable periods of time. Medical practice itself provides the sole and ultimate verdict on the true value of a drug. . . ."

On the other side were Max Finland, the infectious disease specialist from Harvard Medical School who had been one of the early skeptics about Aureomycin; and Harry Dowling, who had been Finland's protégé at Harvard's Thorndike Memorial Library and was, in 1956, chair of the Department of Preventive Medicine at the University of Illinois Medical School. Finland and Dowling, pretty good phrasemakers themselves, sniffed that a physician who chose an antibiotic based on testimonials, rather than peer-reviewed randomized clinical trials, was engaging in "therapeutics by vote."* And, just so no one missed the point that such votes were very easy to rig, Harry Dowling addressed the 1957 annual meeting of the AMA with a much-reprinted speech entitled "Twixt the Cup and the Lip" in which he attacked the use of the techniques "that had been used so successfully in the advertising of soaps and tooth pastes and of cigarettes, automobiles, and whiskey" to market drugs to doctors.

In some ways, the history of antibiotic discovery, from the days of Koch and Pasteur through Paul Ehrlich, Gerhard Domagk, Fleming, Florey, and Hodgkin, can be read as an exercise in epistemology. The great innovations, from Robert Burns Woodward's elegant systems for elucidating chemical structure to Selman Waksman's brute force technique for finding useful soil-dwelling bacteria, were as much about developing new methods for creating knowledge as they were about the knowledge itself. Likewise, no matter how much it looks like a battle over selling one's soul to Pfizer or Merck, the conflict between Welch and Martí-Ibáñez on one side, and Finland and Dowling on the other, was really epistemological: What is the best way to know what actually cures disease?

The significance of this battle to the practice of medicine can hardly be overstated. Several hundred thousand American physicians and at least as many overseas were, for the first time, pushed and

---

* Finland had actually opposed what he viewed as overreliance on RCTs the decade before, arguing that for many diseases in which the etiology was unknown, clinical trials would not necessarily reveal the truth. If physicians don't know whether they're treating bacterial or viral pneumonia, the information from even well-designed clinical trials can be not only useless but harmful, since it fails to control for the messiness of patients with similar symptoms but different diseases.

pulled by two opposing forces: First was their virtually complete dependence on pharmaceutical companies for information about the relative effectiveness of the drugs they prescribed. Second, their no less complete insistence on autonomy in deciding which ones to use. The resulting gap wasn't about information per se, but rather about credibility: Which claims are trustworthy, which not, and, especially, what cognitive tools can be used to decide between the two?

It's tempting to portray the conflict in simple terms. Doctors, after all, remain enormously respected in part because they are at least nominally obliged to place the patient's good above any other consideration, while pharmaceutical companies, however large their contribution to human welfare, remain for-profit enterprises. There are complicating factors, of course. Physicians' practices and hospitals are also businesses, after all; pharmaceutical company researchers are as motivated by the glory of discovery as by their stock options. Even more confounding was and remains the phenomenon that privileges personal knowledge over aggregated evidence, which manifests in doctoring in a particularly acute version. Physicians are famously confident in their clinical experience, inclined to trust the results from a dozen patients they've treated, even in the face of a study examining a thousand patients they've never seen. One possible solution was to evaluate drug effectiveness collectively, rather than individually. But the AMA had, in 1953, stopped issuing its "Seal of Approval" that permitted drugs to be advertised in *JAMA*, and dissolved its at least ostensibly regulatory Council on Pharmacy and Chemistry.

In the face of these cognitive and political challenges, the key epistemological question of the antibiotic revolution—how can we know what medical interventions actually work?—remained unanswered. Because of the increasing complexity of medical decision making, and the cognitive biases that accompany virtually all human behavior, it sometimes seems to be unanswerable. One question that the last century of medicine has answered, though, is how much reliance ought to be placed on the counsel of a single clinician; to epidemiologists and public health researchers, the scariest thing said by a physician is any sentence that begins with, "In my experience."

The hearings that began in Washington, DC, almost exactly eleven months after John Lear's first article in *Saturday Review* were not called in order to solve an epistemological crisis about drug efficacy. As one might guess from the charter of the subcommittee that conducted the hearings—a branch of the Senate Judiciary Committee responsible for oversight on antitrust and monopoly issues—the original intent was to review how pharmaceutical companies priced and marketed their products.

An investigation into drug pricing had been gestating for some time. In 1953, John Blair, a staff economist at the Federal Trade Commission, persuaded his bosses at the FTC to start an inquiry into the business of manufacturing and selling antibiotics. Though Blair was, to put it gently, no friend to large business organizations—in 1938, he had published a polemic entitled *Seeds of Destruction: A Study in the Functional Weaknesses of Capitalism*, a "none-too-happy picture of capitalism and its probable future"—his argument wasn't especially ideological. His own physician and pharmacist had informed him that the price for each of the branded broad-spectrum antibiotics, whether Aureomycin, Terramycin, or Chloromycetin, was identical, and likely to stay that way. The reasons, Blair learned in his preliminary investigation, were the dizzying cross-licensing and cooperative marketing arrangements that followed the tetracycline peace treaty. Parke-Davis, for example, as Blair put it, "sold twenty of [the] fifty-one major drug products included in our study, but produced only one: Chloromycetin."

The business seemed ripe for a broad antitrust investigation, but Blair was unable to convince the commission to proceed. Instead, he took an oblique approach. In 1956, the FTC began work on an exhaustive *Economic Report on Antibiotics Manufacture*. When it was published in June 1958, it revealed that antibiotics had made the pharmaceutical industry the country's most profitable business, with overall profit margins of nearly 11 percent after taxes—twice as much as the average U.S. corporation. For those firms lucky enough to produce broad-spectrum antibiotics, margins were as high as 27 percent.

The seed that had been planted by the OSRD's penicillin project fifteen years before had produced some extremely rich fruit.

The day after the report was issued, the FTC issued a complaint alleging collusion in the marketing of broad-spectrum antibiotics, and questioning the whole structure of the tetracycline patent and cross-licensing agreements. Even so, it took another year, and John Lear's series in the *Saturday Review*, before the subject attracted the attention of anyone outside the bureaucratic world of antitrust regulation.

In September 1959, when Senator Estes Kefauver of Tennessee announced that the Antitrust and Monopoly Subcommittee would hold hearings on America's drug business, he was already one of the country's best-known politicians. He had become a television and newsreel star as the chairman of the Senate Special Committee to Investigate Organized Crime in Interstate Commerce, which riveted the country in 1950 and 1951: Every time a mobster like Frank Costello or Joey Adonis, among the more colorful figures ever to appear before a congressional investigation, invoked Fifth Amendment protections against self-incrimination, the man sitting opposite was the senior senator from the Volunteer State. Kefauver, a candidate for the Democratic presidential nomination in 1952 and its nominee for vice president in 1956, was a New Dealer with impeccable liberal credentials—in 1956, he was one of only three southern senators to refuse to sign the prosegregation Southern Manifesto (the others were Lyndon Johnson and Albert Gore, Sr.)—and a reflexive hostility toward big business of any variety. In short, he was the pharmaceutical industry's worst nightmare: a combination of liberal populism and intellectual range, with an understanding of both old-fashioned political theater and the power of modern media.

Even worse (or better): The first person Kefauver hired to join the subcommittee's staff was John Blair.

On December 7, 1959, in the Old Senate Office Building, the first witnesses were sworn in. For the next ten months, they would arrive, be interrogated in turns by friendly and unfriendly staff and committee members, and depart. As promised, the first subject on offer was pricing: what Kefauver and Blair saw as a shocking distance between the cost of manufacturing a particular drug and its price to consumers. It wasn't just that the prices themselves were what Kefauver thought to be egregiously high. For the Antitrust and Monopoly Subcommittee, what mattered weren't high prices as such, but the suspicion that the prices were being artificially inflated by a conspiracy in restraint of free trade. Since the demand for drugs was determined not by patients, but by a physician's prescription pad, the place to look for such a conspiracy was in the unique marketing practices of the pharmaceutical industry. "The drug industry," as Kefauver put it, "is unusual in that he who buys does not order, and he who orders does not buy."

By 1959, the pharmaceutical industry was developing and marketing considerably more than the antibiotics that had jump-started its growth the preceding decade. As a result, the first witnesses called by the subcommittee gave testimony on other, newer, wonder drugs. One of the first to be examined was the corticosteroid prednisone, an immunosuppressant used to treat diseases like colitis and multiple sclerosis, in which the symptoms are frequently caused by the immune system's own inflammatory response. Francis C. Brown, the president of the Schering Corporation, which introduced the drug in 1955 under the name Meticorten, was blindsided by the initial line of questioning: Why, he was asked, was his company charging some seven hundred times more for a dose of prednisone than it cost to manufacture it? Despite attempts to explain that the price of a drug had to reflect its fixed development expenses as well as its marginal manufacturing cost, Brown had already lost the public relations battle. The front page of the *New York Times* for December 8 read, "SENATE PANEL CITES MARK-UP ON DRUGS RANGING TO 7,079%."

And so it went, for month after month. The subcommittee's chief counsel, Rand Dixon, stated for the record that the Upjohn Company used only 14 cents worth of raw materials to make a drug it sold for $15, a markup of "about 10,000 percent." Corticosteroids gave way to tranquilizers—the term had only recently been coined to describe Miltown, the brand name for the mild sedative meprobamate, and the world's first blockbuster psychotropic drug—to arthritis medications, to antidiabetic drugs. Physicians and hospital directors went on record accusing pharmaceutical firms of "brainwashing" tactics and "perverted marketing attitudes."

Meanwhile, the subcommittee's Republicans, led by Senator Everett Dirksen of Illinois,* counterpunched, asking why, if pharmaceutical companies were marking up prices several thousandfold, they were still only managing to achieve profit margins of less than 15 percent.

By the spring of 1960, however, the subcommittee's concerns had expanded from pricing and marketing strategies to patent and trademark reform, particularly regarding the ways in which branded drugs—especially the proprietary fixed-dose combination drugs like Sigmamycin—were being offered the same sort of intellectual property protections as compounds like streptomycin, even though they weren't required to show that they were truly novel or even effective. The stage was set for the main event: antibiotics.

The stars of the hearings' climax ought to have been Henry Welch and Félix Martí-Ibáñez, both of whom had demanded an opportunity to appear and clear their names. Kefauver had taken the two up on their offers, notifying them that they would be given a chance to do so at the hearing scheduled for May 17, 1960. Neither appeared, pleading illnesses that stubbornly, and suspiciously, hung on until the hearings ended in September.

Their absence had little impact on the hearings' theatrics. At one point, Rand Dixon asked Dr. Perrin Long, a pioneer in the use of sulfa drugs, "Do you think a cost of $17 [for antibiotics] to the average mother or father every time their child has a cold is down to a point where it can be reached even by the needy?" The question made headlines; Dr. Long's most relevant response—that an antibiotic would be useless at any price for treating a cold—went unsaid. Harry Loynd of Parke-Davis, called to account for both the marginal cost for the active ingredient in Chloromycetin *and* the battle over the way it had been promoted during the aplastic anemia scare, was a perfect foil for Kefauver. Loynd had never learned to hide his contempt for politicians,

* Dirksen described Kefauver as combining the charm of a Victorian lady with the single-mindedness of an Apache.

and came off not as a no-nonsense executive without the time to suffer fools gladly, but as an arrogant stuffed shirt. And, worse, an evasive one, who fought with Kefauver over every comma in every document, even to the point of arguing whether he had "seen" or simply "been aware" of an ad for Chloromycetin that seemed to underplay its risks.

The twenty-one months that began with John Lear's first article for the *Saturday Review* on January 3, 1959, and ended with the close of the Kefauver hearings on September 14, 1960, were as earthshaking, in their own way, as the two years following the transatlantic trip of Howard Florey and Norman Heatley in 1941. Though specific drugs and pharmaceutical companies had for years been the objects of criticism from academic physicians like Maxwell Finland and gadflies like Edgar Elfstrom, in 1959, the public at large remained almost uniformly optimistic about the era of the wonder drugs, in which miracles— from penicillin to the Salk vaccine—had appeared, it seemed, almost every day. By the end of 1960, they would never be quite so enthusiastic again. The most dispiriting news about the wonder drugs, it turned out, wasn't that they were overpriced; it was that no one knew whether they were really effective. Haskell Weinstein, formerly the medical director of one of Pfizer's subsidiaries, had revealed the "very prevalent misconception" that the FDA was required to document efficacy. "As a physician I blush with shame at the quality of some of the 'studies' done by some of my physician brethren."

When the Kefauver hearings closed in September 1960, the subcommittee had remarkably little to say about drug pricing or pharmaceutical monopolies. It recommended, instead, expanding the statutory authority of the FDA, which had largely been frozen since 1938. Henceforth, the agency should require proof of efficacy, as well as safety, of all new drugs, and "apply certification procedures to all antimicrobial agents for use in infectious diseases." In April 1961, Kefauver introduced Senate Bill 1522 along essentially the same lines.

Testimony on the bill would occupy the next seven months. The FDA was—no surprise—a big supporter. More surprising: So were many of the largest pharmaceutical firms, though not Parke-Davis;

Harry Loynd was still smarting from his treatment by the senior senator from Tennessee. On the other hand, the American Medical Association opposed Kefauver's bill most strenuously because their membership was violently hostile to the idea that drug efficacy could be known by anyone other than the individual physician. Unstated was the concern that the restrictions on pharmaceutical advertising that Kefauver had included in the bill would fall most heavily on *JAMA*.*

Despite widespread public endorsements, the support of strong majorities in both houses of Congress, and even the pharmaceutical industry itself, SB1522 seemed doomed to die the death of a thousand cuts in committee. And so it might have, but for a scandal more gruesome, and more notorious, than the Elixir Sulfanilamide and aplastic anemia scares combined.

In 1962, Frances Oldham Kelsey had been balancing the costs and benefits of the antibiotic revolution for nearly twenty-five years. In 1938, with the ink barely dry on her doctorate in pharmacology from the University of Chicago, she performed the animal studies that revealed the extent of the damage caused by Massengill's Elixir Sulfanilamide. Twelve years later, she became an MD as well as a PhD; and, in August 1960, Dr. Kelsey joined the Food and Drug Administration as one of only seven full-time drug reviewers. Her first assignment was an application from Richardson-Merrell, Inc., a drug wholesaler in Cincinnati then best known for the menthol-infused petrolatum known as Vicks VapoRub. The drug for which the company had applied for U.S. marketing rights, to be called Kevadon, had achieved enormous popularity in western Europe as a sedative: one that was safer than barbiturates, and, in addition, was effective

* Worth noting: Austin Smith, the former editor of *JAMA*, had become president of the Pharmaceutical Manufacturers Association . . . and would succeed Harry Loynd as president of Parke-Davis.

as an antinausea drug. The German pharmaceutical firm Chemie Grünenthal—a postwar company that got its start selling penicillin under the Allied occupation—had developed and sold it, first by prescription, then direct to consumers, as Contergan. In the United Kingdom, Distillers Limited marketed it as Distaval. Its generic name was thalidomide.

In 1960, pharmaceutical companies were able to contact FDA reviewers directly as often as they pleased, and Richardson-Merrell pressed Dr. Kelsey for a quick approval to their application, which they expected in a matter of months, given the widespread use of the drug in Europe. In fact, the FDA automatically granted approval for new drugs if a reviewer failed to act on applications within sixty days . . . but action was understood to include any request for additional information. And Frances Kelsey definitely wanted more information. She asked for clinical and animal studies on toxicity. She requested data on the drug's effect on pregnant women, given that it was being touted as a safe cure for morning sickness. Richardson-Merrell supplied testimonials. They sent more letters, and what purported to be the best research on the new product. Kelsey called it "an interesting collection of meaningless pseudoscientific jargon apparently intended to impress chemically unsophisticated readers." Drug company representatives visited Kelsey in her office dozens of times. Richardson-Merrell's executives went over Kelsey's head, to George Larrick, the commissioner of Food and Drugs. Larrick—the inspector who had mobilized the entire FDA field force to track down virtually every drop of Elixir Sulfanilamide in 1937—backed his reviewer. Every sixty days, Frances Kelsey sent another letter informing Richardson-Merrell that their application still awaited approval.

And so it went, until November 29, 1961, when Chemie Grünenthal sent Richardson-Merrell the first reports of phocomelia—"seal limb," a birth defect that caused stunted arms and legs, fused fingers and thumbs, and death; mortality rates for the condition approached 50 percent. Phocomelia wasn't, in the 1960s, unknown. But it had been an extremely rare genetic disorder, with fewer than a thousand reported cases *worldwide*. No longer. Eight West German pediatric clinics had reported no cases of phocomelia between 1954 and 1959. In 1959, they reported 12. In 1960, there were 83. In 1961, 302. No one needed a sensitive statistical test to tease out the cause.* The mothers of the malformed infants had all taken thalidomide.

By the time the drug was removed from sale at the end of 1961, hundreds of thalidomide babies were struggling for life. Just as horrifying: Tens of thousands of expectant mothers who had taken the sedative spent the last months of their pregnancies consumed by a completely rational fear of how they would end. By the time the last exposed mothers gave birth, the total number of phocomelic infants exceeded ten thousand. Thanks entirely to Frances Kelsey's stubbornness, fewer than thirty of them had been born in the United States.

The reason for even that small number was that Richardson-Merrell had recruited physicians for "investigational use" of the drug prior to FDA approval, which was not only permissible but condoned by the existing 1938 Food, Drug, and Cosmetic Act. As a result, when the company withdrew its application at the end of 1961, the long tail of thalidomide risk hadn't yet been reached. Kelsey, very much aware of this, sent the company a letter asking whether any quantity of Kevadon/thalidomide was still in the hands of physicians. The company was unable to provide anything but an embarrassingly incomplete answer; it had distributed more than 2.5 million thalidomide pills to more than a thousand doctors in the United States and had utterly failed to maintain adequate records of who, when, and how much. Most of the expectant mothers in the United States who had been given the sedative by their physicians hadn't even been told that the drug was experimental.

---

* The mechanism by which thalidomide causes birth defects is still not very well understood. A currently popular theory is that one of the simpler compounds into which the liver breaks down the drug inhibits the formation of blood vessels at the end of embryonic development, and so damages the growth of structures that appear late in gestation, such as arms and legs. The drug remains in use today, for treatment of—among other things—leprosy and multiple myeloma; a reminder that drug "safety" is very much a relative thing.

Despite the tragic stories of victims, and the embarrassing revelations about the holes in the approval process—in 1960 alone, the FDA had received thousands of applications, and it was only by great good luck that Frances Kelsey was the one to whom the Kevadon application had been assigned—thalidomide didn't really become a scandal until July 15, 1962, when Morton Mintz of the *Washington Post* published a front-page story, with the headline: "'HEROINE' OF FDA KEEPS BAD DRUG OFF MARKET." Its first sentence read:

> *This is the story of how the skepticism and stubbornness of a Government physician prevented what could have been an appalling American tragedy, the birth of hundreds or indeed thousands of armless and legless children.*

The *Post* story generated hundreds of comments and opinion pieces throughout the country. On August 8, 1962, Frances Kelsey was honored with the President's Award for Distinguished Federal Civilian Service; in the words of Senator Kefauver, she had exhibited "a rare combination of factors: a knowledge of medicine, a knowledge of pharmacology, a keen intellect and inquiring mind, the imagination to connect apparently isolated bits of information, and the strength of character to resist strong pressures." Within weeks, SB1522 was taken off life support, and on August 23 the House and Senate passed the Kefauver-Harris Amendments (the bill had been introduced in the House of Representatives by Oren Harris of Arkansas). On October 10, 1962, Public Law 87-781, an "Act to protect the public health by amending the Federal Food, Drug, and Cosmetic Act to assure the safety, effectiveness, and reliability of drugs," was signed into law by President John Kennedy. Standing behind him for the traditional signing photo was Frances Oldham Kelsey.

Credit: National Institutes of Health/National Library of Medicine

Frances Oldham Kelsey (1914–2015) receiving the President's Award for
Distinguished Federal Civilian Service from President John F. Kennedy

Kefauver-Harris wasn't the first major piece of federal legislation to recognize that the world of medicine had been utterly transformed since
1938. In 1951, Senator Hubert Humphrey of Minnesota and Represen-
tative Carl Durham of North Carolina—both, not at all coincidentally,
had been pharmacists before entering political life—cosponsored
another amendment that drew, for the first time, a clear distinction between prescription drugs and those sold directly to patients.

Until the 1950s, the decision to classify a drug as either a prescription drug, requiring physician authorization, or as what is now known
as an over-the-counter medication, was entirely at the discretion of

the drug's manufacturer. This was one of the longer-lasting corollaries of the nineteenth-century principle that, because of the sanctity of consumer choice, people had an inalienable right to self-medicate. As a result, the decision to classify a drug as prescription only was just as likely to be made for marketing advantage as safety considerations: An American drug company could, and did, decide that prices could be higher on compounds that were sanctioned by physicians. Predictably, therefore, the same compound that Squibb made available by prescription only could be sold over-the-counter by Parke-Davis.

After Humphrey-Durham, any drug that was believed by the FDA to be dangerous enough to require supervision or likely to be habit forming, or any new drug approved under the safety provision of the 1938 act, would be available only by prescription; further, the drug and any refills were required to carry the statement "Federal law prohibits dispensing without prescription." All drugs that could be sold directly to consumers, on the other hand, had to include adequate directions for use and appropriate warnings, which is why even a bottle of ibuprofen tells users to be on the lookout for the symptoms of stomach bleeding.

Humphrey-Durham was intended to protect pharmacists from prosecution for violating the many conflicting and ambiguous laws about dispensing drugs. By the end of the 1940s, American pharmaceutical companies were selling more than 1,500 barbiturates, all basically the same, but the regulations governing them barely deserved to be called a patchwork. Thirty-six states required prescriptions; twelve didn't. Fifteen either prohibited refills or allowed them only with a prescription. And while some pharmacists viewed this as a loophole through which carloads of pills could be driven—one drugstore in Waco, Texas, dispensed more than 45,000 doses of Nembutal, none of them by prescription—others were arrested for what amounted to little more than poor record keeping. Even where pharmacies weren't attempting to narcotize entire cities, risks didn't vanish. A Kansas City woman refilled her original prescription (for ten barbiturate pills) forty-three times at a dozen different pharmacies before she was discovered in her home, dead, partially eaten by rats.

In its original draft, the sponsors of Humphrey-Durham had tried to provide more than just a "clear-cut method of distinguishing be-tween 'prescription drugs' . . . and 'over-the-counter drugs.'" In the second draft of the bill, the FDA administrator, "on the basis of opin-ions generally held among experts qualified by scientific training and experience to evaluate the safety and efficacy of such drug," was charged with deciding whether a drug was unsafe or ineffective with-out professional supervision.

By the time the amendment was signed, however, any language about effectiveness had been negotiated away. In 1951, there was no constituency, either among patients, physicians, or pharmaceutical com-panies, urging the FDA to evaluate effectiveness. However, what had been untenable in 1951 became the law of the land in 1962. New drugs, finally, would have to be certified not merely as safe, but as effective.

The new law didn't restrict itself to new compounds. The 1962 amendment also required the FDA to review every drug that had been introduced between 1938 and 1963 and assign it to one of six categories: effective; probably effective; possibly effective; effective but not for all recommended uses; ineffective as a fixed combination; and ineffective. Chloramphenicol, for example, was designated as "probably effective" for meningeal infections, "possibly effective" for treatment of staph infections, and, because of the risk of aplastic ane-mia, "effective but . . ." for rickettsial diseases like typhoid fever and plague. The review process, known as DESI (for Drug Efficacy Study Implementation), began in 1966, when the FDA contracted with the National Research Council to evaluate four thousand of the sixteen thousand drugs that the agency had certified as safe between 1938 and 1962.* Nearly three hundred were removed from the market.

In 1963, Frances Kelsey was named to run one of five new branches in the FDA's Division of New Drugs, the Investigational Drug Branch (now known as the Office of Scientific Investigations). She was tasked

* A large number of over-the-counter drugs were designated by the acronym GRASE (for Generally Recognized as Safe and Effective). Others that had appeared before 1938 were grandfathered . . . though hardly any qualify any longer, since one provi-sion of the Kefauver-Harris Amendments required that it be identical to the version on sale in 1962.

with turning the vague language of the Kefauver-Harris Amendments into a rule book. The explicit requirements of the law weren't actually all that explicit. It required "substantial evidence" of effectiveness that relied on "adequate and well-controlled studies" without actually defining the term. Like the 1938 act, which called for only "adequate tests by all methods reasonably applicable," the amendments didn't specify any particular criterion for evaluating either safety or efficacy. It was a statement of goals, not strategies.

Determining which strategies would be most effective was the next step. Though Bradford Hill's streptomycin trials of 1946 had demonstrated the immense hypothesis-testing value of properly designed randomized experiments, ten years later nearly half of the so-called clinical trials being performed in the United States and Britain still didn't even have control groups. Though one pharmaceutical company executive after another had appeared before the Kefauver investigators to claim that the huge sums invested in clinical research justified high drug prices, they were spending virtually all of their research dollars on the front end of the process: finding likely sources for antibiotics, for example, then extracting, purifying, synthesizing, and manufacturing them. The resources devoted to discovering whether they actually worked outside the lab were minuscule by comparison: essentially giving away free samples to physicians and collecting reports of their experience. As Dr. Louis Lasagna, head of the Department of Clinical Pharmacology at Johns Hopkins, had told the Kefauver committee, controlled comparisons of drugs were "almost impossible to find."

Frances Kelsey wasn't any more inclined to accept the status quo than she was to believe the "meaningless pseudoscientific jargon" that Richardson-Merrell had offered in support of their thalidomide application. In January 1963, even before she was named to head the Investigational Drug Branch, Kelsey presented a protocol for reviewing what was now termed an "Investigational New Drug." The new system would require applicants for FDA approval to present a substantial dossier on any new drug along with their initial application. Each IND, in Kelsey's proposed system, would need to provide information on animal testing, for example—not just toxicity, but effectiveness. Pharmaceutical companies would be obliged to share information about the proposed manufacturing process, and about the chemical mechanism by which they believed the new drug offered a therapeutic benefit. And, before any human tests could begin, applicants would have to guarantee that an independent committee at each institution where the drug was to be studied would certify that the study was likely to have more benefits than risks; that any distress for experimental subjects would be minimized; and that all participants gave what was just starting to be known as "informed consent."*

The truly radical transformation, however, was what the FDA would demand of the studies themselves. Kelsey's new system specified three sequential investigative stages for any new drug. The first, phase 1 clinical trials, would be used to determine human toxicity by providing escalating doses to a few dozen subjects in order to establish a safe dosage range. Compounds that survived phase 1 would then be tested on a few hundred subjects in a phase 2 clinical trial, intended to discover whether the drug's therapeutic effect—if any— could be shown, statistically, to be more than pure chance. The final hurdle set out by the 1963 regulation, a phase 3 trial, would establish the new drug's value in clinical practice: its safety, effectiveness, and optimum dosage schedules. Phase 3 trials would, therefore, require larger groups, generally a few thousand subjects, tested at multiple locations. At the latter two stages, but especially the third, the FDA gave priority to studies that featured randomization, along with experimental and control "arms." If the new drug was intended to treat a condition for which no standard treatment yet existed, it could be compared ethically against a placebo. If, as was already the case for most infections and an increasing number of other diseases, a treatment already existed, studies would be obliged to test for "non-inferiority," which is just what it sounds like: whether the effectiveness of the new treatment isn't demonstrably inferior to an existing one. In either case, the reviewers at the FDA would be far more likely to

---

* The first use of the term seems to be no older than 1957, when it formed part of the argument in a medical malpractice case.

grant approval if the two arms in an approved study were double-blinded, with neither the investigators nor subjects aware of who was in the experimental or control groups.

In February 1963, the commissioner of Food and Drugs approved Kelsey's three-tiered structure for clinical trials. The process of pharmaceutical development would never be the same. It marked an immediate, though temporary, shift of power from pharmaceutical companies to federal regulators. Within weeks of the announcement of the new regulations, virtually every drug trial in the country, from the Mayo Clinic to the smallest pharmaceutical company, was reclassified into one of the three allowable phases. It allowed Frances Kelsey a remarkably free hand in exercising her authority to grant or withhold IND status; to her critics, this led to any number of cases in which she withheld classification based on nothing but a lack of faith in a particular investigator, or her judgment that the proposed drug was either ineffective or dangerous.*

The new requirements, which would remain largely unchanged for at least the next fifty years, permanently altered the character of medical innovation.

The method of validating medical innovation using randomized control trials had given the world of medicine a way of identifying the sort of treatments whose curative powers weren't immediately obvious to clinicians (and, just as important, identifying those that seemed spectacular, but weren't). Until 1963, however, RCTs had been a choice. The three phases of the newly empowered FDA made them a de facto requirement. Frances Kelsey's intention was to use the objectivity of clinical trials to simultaneously protect the public and promote innovative therapies. It is unclear whether she understood the price.

One of the underrated aspects of the wave of technological innovation that began with the first steam engines in the eighteenth century—the period known as the Industrial Revolution—was a newfound ability

---

* Her instinct to lecture drug companies on moral as well as scientific grounds soon made her as notorious among pharmaceutical companies as she was lionized by the general public.

to measure the costs and benefits of even tiny improvements, and so make invention sustainable. Just as improvements in the first fossil-fueled machines could be evaluated by balancing the amount of work they did with the amount of fuel they burned, even small benefits of new drugs and other therapies could be judged using the techniques of double-blinding and randomization. Since almost all potential improvements are by definition small, medicine generally, and the pharmaceutical industry in particular, now had a method for sustaining innovation. No longer would progress wait on uncertain bursts of genius; discovery could now be systematized and even industrialized.

However, there was a giant difference between the methods used to compare mechanical inventions and medical or pharmaceutical treatments. Engineers don't need to try a new valve on a hundred thousand different pumps to see whether it improves on an existing design. But so long as the RCT was the gold standard for measuring improvement in a drug (or any health technology), small improvements in efficacy would require larger, more time-consuming, and costlier trials. By the arithmetic alone, the value of a treatment that is so superior to its predecessor that it saves ten times more people is apparent after only a few dozen tests. One that saves 5 percent more can require thousands. The smaller the improvement, the more expensive the testing would become.

This changed the calculus of discovery dramatically. Selman Waksman's technique for finding new drugs—sifting through thousands of potential candidates in order to find a single winner—had already virtually destroyed the belief that a brilliant or lucky scientist, working alone (or, more likely, in a relatively small laboratory in a university or hospital), might find a promising new molecule. But *demonstrating that it worked* would, thanks to Frances Kelsey and Bradford Hill, make the process exponentially more expensive, and riskier. The same economies of scale that had been necessary for the manufacture of the first antibiotics were now required for finding and testing all the ones that would follow. Perversely, the Kefauver hearings, initiated and stage-managed by liberal politicians with no love for big business, had led inexorably to the creation of one of the largest and most profitable industries on the planet.

Engineers calculate failure rates—sometimes they're known as "failure densities"—to describe phenomena like the increasing probability over time that one of the components of an engine drivetrain will crack up. Pension companies use similar-looking equations to calculate life spans. Medical researchers use them to derive the survival probabilities of patients given different treatment regimens.

Another way of applying the arithmetic of failure rates is to use it to predict the number of promising "new molecular entities" that will actually prove out as therapeutically useful. A pharmaceutical company that identifies a thousand compounds with some potential for the treatment of a disease like Alzheimer's, for example, and knows that the failure rate in preliminary testing was somewhere between 95 and 99 percent, could guess that between ten and fifty compounds might survive to the next round. Moreover, if the failure rate were constant, then the likelihood of success would increase over time; the longer you spend looking for the next miracle drug, the closer you are to finding it.

The relevance for the riskiness of drug development is fairly clear. If pharmaceutical research were characterized by a constant failure rate—even a very high one—it might be expensive, but not particularly risky: At any given moment, the probability of a successful outcome would be known. If, on the other hand, the failure rate were fundamentally variable, then years will be spent on completely fruitless searching.

Drug development, from proof of concept (sometimes called "phase 0") through phase 3 clinical trials, has never exhibited anything resembling a constant failure rate. This means that it has inevitably grown riskier over time. Many close observers of the phenomenon have argued that this is a reason for transferring the risk of pharmaceutical innovation to society at large, by increasing government support both for basic biomedical research and for testing the products of that research.

There's no intrinsic reason why government agencies (or not-for-profit institutions like universities) are fundamentally incapable of funding, or even managing, the phased testing system that Frances

Kelsey developed at the FDA more than fifty years ago. However, their involvement would not fundamentally alter the relationship between risks and rewards. For more than a century, society has farmed out the risk of pharmaceutical development, testing, and manufacturing to the institutions willing to undertake it, but they only do so when the potential rewards are large. Inveighing against pharmaceutical company greed just camouflages this unavoidable truth.

The machine of pharmaceutical innovation—one that wouldn't exist, would never even have been built, but for the antibiotic revolution—is decidedly imperfect, full of inefficiencies and side effects, and incredibly costly to run and maintain. Paul Ehrlich's side chains, or Dorothy Crowfoot Hodgkin's X-ray crystallography, or even Norman Heatley's jury-rigged distillation apparatus were the result of motivated intellectual effort in search of some reward. The motivations haven't always been particularly noble; Pasteur's hatred of Germany for the Franco-Prussian War and Ernst Chain's lust for a Nobel Prize come to mind. Only a very jaundiced observer, though, would think the bargain wholly bad. Winston Churchill famously observed, "It's been said that democracy is the worst form of Government except for all those other forms that have been tried from time to time. . . ."

The pharmaceutical industry is rightly criticized for spending millions to acquire knowledge that it then uses to find more expensive treatments for existing conditions. Or even to "medicalize" conditions in order to create a market for a new treatment.

On the other hand, consider HAART.

Though the disease was first discovered in 1981 (and given the name AIDS a year later), the virus responsible wasn't identified until 1983. Even before then, Burroughs Wellcome's U.S. subsidiary, with its long history of investigating obscure diseases, was researching the new and horrifying disease that killed its hosts by destroying their immune systems and so exposing them to hitherto rare syndromes like the cancer known as Kaposi's sarcoma and, even more relevant, retroviruses like HIV. In 1983, one of the company's biochemists, Jane

Rideout, started investigating the chemical properties of an antibacterial compound known as azidothymidine: AZT for short.

Simultaneously, other researchers at Burroughs Wellcome had pioneered a dramatically different way of testing new molecules for effectiveness; instead of Selman Waksman's time-honored trial-and-error method of testing large numbers of promising chemical compounds, the new technique—which would win a Nobel Prize for its inventors—required identifying a chemical that the target pathogen needs to reproduce, and replacing it with an analogue that attracts the pathogen while sabotaging its reproduction. Rideout realized that AZT was a near-perfect analogue for a chemical required for HIV . . . and Burroughs Wellcome agreed. Only three years after the human immunodeficiency virus was first identified—a pace that recalls the original antibiotic revolution—the Burroughs Wellcome company introduced AZT as the first effective anti-AIDS drug.* A decade later, when Merck received FDA approval for the antiviral drug Indinavir, and a separate set of approvals were granted to drugs known as NRTIs (nucleoside reverse transcriptase inhibitors), the combination therapy known as HAART, for highly active antiretroviral therapy, transformed HIV from a death sentence to a chronic, and treatable, condition.

HIV-positive patients are scarcely alone in their debt to pharmaceutical innovation. Tens, perhaps hundreds, of millions of victims of a thousand diseases from leukemia to river blindness are alive and thriving entirely because of a drug breakthrough. For them, and especially for the literally uncountable number of people whose bacterial infections, from strep throat to typhus to anthrax, were cured by a ten-day regimen of antibiotics, the bargain probably seems an extraordinarily one-sided one. Like Anne Miller and Patricia Thomas, they were, and are, living, breathing evidence that Joseph Lister's dream came true.

---

* Not without controversy. When Burroughs announced the initial price for a year's treatment with AZT—a then-unthinkable $10,000—they were vilified in the press as profiteering monsters.

EPILOGUE

"The Adaptability of the Chemist"

Many of the institutions at the forefront of research on infectious disease are as prominent today as they were during the years of the antibiotic revolution, and even before. While the Lister Institute of Preventive Medicine is now a charity employing fewer than half a dozen full-time staff members, it continues to fund research and disburse prize fellowships, some of them very substantial. The Robert Koch Institute is an agency of the Federal Republic of Germany, still performing significant research, though with more emphasis on epidemiology than pharmacology or biochemistry. The Institut Pasteur remains one of the world's preeminent research institutions, with more than a hundred labs and a thousand scientists working throughout the world on basic biochemistry, virology, and a huge number of other subjects.

Oxford's Sir William Dunn School of Pathology continues to occupy the cutting edge of research into human health and disease, publishing papers on everything from T-cell activation to intercellular signaling to—yes—bacterial pathogenesis. And St. Mary's Hospital still runs a Centre for Infection Prevention and Management (and is custodian for Alexander Fleming's laboratory, now a museum).

Peoria's Northern Lab—renamed the National Center for Agricultural Utilization Research in 1990, but still known to almost everyone as the Northern Lab—remains committed to producing more and better agricultural produce, while also maintaining the Agricultural Research Service's Microbial Culture Collection, nearly a hundred thousand strains of actinomycetes, yeasts, and molds that are made available to research institutions throughout the world. In 1965, the Rockefeller

Institute for Medical Research became Rockefeller University, but didn't miss a step in any other sense; twelve Nobel Prizes in Physiology or Medicine were awarded to its researchers over the next fifty years.

The laboratories built by Abbott, Squibb, Merck, Eli Lilly, and others in the 1930s continue as factories of innovation, though the drug industry has experienced a vertigo-inducing series of corporate makeovers since the beginning of the antibiotic revolution. Some of the original companies recruited by A. N. Richards and the OSRD in 1943 are still recognizable, though each is enormously larger. Merck, which merged in 1953 with the Philadelphia-based Sharp & Dohme and subsequently acquired half a dozen other competitors, including Schering-Plough, now produces revenue in excess of $40 billion annually. Pfizer is even bigger, a company with sales of more than $50 billion. Eli Lilly is a $23-billion company. The combination of Bristol-Myers and Squibb, which merged in 1989, weighs in at nearly $20 billion as does Abbott Laboratories.

Others are no longer going concerns, run onto the rocks by waves of the "creative destruction" that the Austrian economist Joseph Schumpeter called the defining characteristic of capitalism. In 1988, Eastman Kodak acquired Winthrop (or Sterling Winthrop), a member of the original penicillin project *and* the discoverer of the first quinolone antibiotics. It was then broken apart and sold, in pieces: to the French pharmaceutical company Sanofi, to the British firm Smith-Kline Beecham (a successor to the original Beecham's Pills, now known as GlaxoSmithKline), and to the revived German giant, Bayer, which, as a result, finally reacquired the rights to the name "Bayer Aspirin." Earlier, in 1974, Bayer had acquired Cutter Laboratories. Parke-Davis, America's biggest pharmaceutical company through the 1950s, never achieved that status again; in 1976, it was acquired by Warner-Lambert. When Pfizer acquired Warner-Lambert in 1976, though, the most valuable asset in the transaction was a discovery made in Parke-Davis's labs: the cholesterol-reducer atorvastatin, which, as Lipitor, became the most profitable drug of all time. Hoechst AG, where Emil Behring and Paul Ehrlich made history, was reconstituted after the Second World War. In 1999, it merged with Rhône-Poulenc (the onetime employer of German pharmacology's nemesis, Ernest Fourneau). In 2004, the combined company—briefly Aventis—was acquired by Sanofi.

It is difficult to calculate, with all those acquisitions and divestitures, just how large the enterprises borne out of the original penicillin project became. A reasonable guess is that they deliver to their shareholders somewhere north of $40 billion in operating income annually. What they don't deliver much of is antibiotics. Though Pfizer still makes four antibacterial compounds, in 2011 the company closed its dedicated-to-antibiotics Connecticut research lab. Roche, Bristol-Myers Squibb, and Eli Lilly—all charter members of the penicillin project—no longer make any antibiotics at all. Neither does Johnson & Johnson, the largest pharmaceutical company in the world.

The primary reason is that it's extraordinarily difficult to find new antibiotics. After sixty years, almost every antibiotic that remains on pharmacy shelves still uses one of a very limited number of methods for attacking pathogens—disrupting bacterial DNA, weakening bacterial cell walls, inhibiting the enzymes used by bacteria to synthesize proteins—that were used by the original beta-lactams, macrolides, and tetracyclines. The successors to penicillin and erythromycin are more effective and less toxic than the versions that started the antibiotic revolution in the 1940s, but they're refined versions of a seventy-year-old biochemical technology. Molecules aimed at new targets, such as drugs that disrupt bacterial DNA synthesis (by, for example, inhibiting the enzyme that allows DNA to unwind without breaking), are regularly tested. A few have made it all the way into clinical trials.

In addition, almost all of the newly discovered molecules that show some antibacterial potential have the tyrothricin problem: They're just as toxic to humans as they are to pathogens, which places something of a ceiling on their appeal. It's because antibiotic-resistant infections have become so dangerous, and so ubiquitous, that drugs like colistin, first isolated in 1949 but so toxic to kidneys and the nervous system that it never came into wide use, is now a last-resort drug for resistant Gram-negative infections. When a patient is at risk of death from an infection that is resistant to safer antibiotics, the risk of kidney failure appears less daunting.

The genomic revolution, by identifying which genes were the blueprints for essential proteins, could, in theory, have empowered medicinal chemists with the ability to target only the genes necessary for bacterial survival and leave the ones for mammals untouched. It's easy to see why this seemed so promising; knowing every aspect of a particular bacterium's genetic makeup—what it eats, how it reproduces—would surely produce true "magic bullets."

The promise of genomic antibacterials remains unfulfilled. The first bacterial genome was sequenced in 1995—*Haemophilus influenzae*, the likely killer of George Washington in 1799—and thousands of genes were identified shortly thereafter as potential antibacterial targets, because they produced proteins essential to bacterial survival. Dozens of pharmaceutical companies evaluated them, exposing the genes to literally hundreds of thousands of molecular compounds (GlaxoSmithKline, between 1995 and 2001, assayed nearly half a million alone). Seven years, and more than $100 million later, fewer than half a dozen even qualified as "hits": potential "lead molecules." Given historical rates of attrition, the number that might even make it to the next step—as a "development candidate"—is statistically indistinguishable from zero.

Even if weren't so difficult to find new antibiotics, their very nature makes them a suboptimal long-term investment given the need to allocate limited resources among a wide range of alternatives. Antibiotics were, in a sense, victims of their own dramatic effectiveness. A drug that does its job in ten days can't possibly compete for institutional resources with one that will be taken every day for a lifetime. The managers and shareholders of companies like Merck, Pfizer, and Eli Lilly didn't require very sophisticated arithmetic to see a greater potential return from drugs that treated chronic ailments rather than acute infections. And they invested their research and validation assets accordingly. From 1962, when George Lesher of Sterling

Winthrop* discovered the first of the quinolone antibiotics, until 2000 not a single new class of antibacterial drugs appeared. Between 2011 and 2013, the FDA approved only three new molecular compounds that *might* combat bacterial pathogens. The cost of developing a new antibiotic is higher, in relative terms, than the price for a new drug to treat depression, or cancer, or hypertension.

What are those costs? The most widely cited method for calculating the cost of drug development estimated, in 2003, that the average out-of-pocket cost for a drug at the moment it received marketing approval from the FDA was more than $400 million. Another calculation, made in 2011, and no less controversial, came up with a median R & D cost of $43.4 million.

There are two primary reasons for the huge difference between the estimates, each of which is regularly used as a bludgeon in the never-ending debate over drug prices. The lower number fails to account for any costs incurred prior to the submission of the drug for the first stage of FDA approval; this "phase 0" stage, in which thousands of potential molecules are screened for evidence of antibacterial activity, and hundreds extracted in quantities sufficient for testing, can take more than five years, and is responsible for fully 30 percent of the $400 million figure. The lowball calculation only estimates the R & D costs directly attributable to the new and approved drug. Since most of the R & D budget of a large pharmaceutical firm is spent on drugs that never make it to market, failing to account for those dollars somewhere is a fairly significant bit of financial sleight of hand. A number that estimates total R & D costs without any basic research costs because—as the author of the original paper wrote—"there is no reasonable estimate available," doesn't encourage much faith in the number's precision.

This doesn't mean that the figures provided by drug companies themselves are disinterested and therefore reliably accurate. The big pharmaceutical corporations are regularly accused of profiteering; of

---

* Coming full circle: Winthrop Chemical Company, in partnership with Bayer and I. G. Farben, was the first company to sell sulfanilamide in the United States.

overcharging for lifesaving medications; of financing favorable research and suppressing negative results; of producing drugs to treat conditions that are virtually nonexistent or new and expensive versions of drugs that are no better than the ones already on offer. Companies with those sort of image problems have every reason to magnify the size of their research budgets, if only to slow their descent in public esteem.

However, the commitment of pharmaceutical companies to research isn't just a PR strategy. Pfizer's audited spending on research and development exceeds $11 billion a year. If it really cost less than $50 million to bring a new drug to market—if the much-cited $43.4 million figure were accurate—this would suggest that Pfizer alone should be launching 250 new drugs annually.

In 2012, the best year for new drug approvals since 1996, the FDA approved a total of thirty-seven "new molecular entities" for the entire pharmaceutical industry.

None of them were antibiotics.

If you leave the Smithsonian Museum of American History, the custodian of Anne Miller's world-historic medical chart, and walk two blocks east along Constitution Avenue until you reach Fifteenth Street, then turn right and proceed for half a mile until you reach Pennsylvania Avenue, you'll find yourself facing the White House. There, on July 9, 2012, seventy years after penicillin pulled Mrs. Miller back from the brink of death, President Barack Obama signed into law Senate Bill 1387: the Food and Drug Administration Safety and Innovation Act. The new law, yet another amendment to the original 1938 Food, Drug, and Cosmetic Act, included dozens of provisions regarding everything from a new protocol for the approval of medical devices, to the authorization of fees for users of generic and prescription pharmaceuticals, to protection of the global supply chain for finished drugs.

Less well publicized, but almost certainly as important, it incorporated into the legislation a series of provisions known collectively by the acronym GAIN: Generating Antibiotic Incentives Now. Intended to increase the likelihood that pharmaceutical companies would invest in the development of antibiotics that treat serious or life-threatening conditions, GAIN offered fast-track approval for compounds that promised to combat infections, and a five-year extension of their exclusive term of patent.

The reason for GAIN and other similar proposals was simple and terrifying. In the United States alone, antibiotic-resistant bacteria now infect two million people annually. More than twenty thousand of them die. Alexander Fleming's observation in his 1945 Nobel Lecture—"It is not difficult to make microbes resistant to penicillin in the laboratory by exposing them to concentrations not sufficient to kill them"—had been an understatement of massive proportions. Infectious disease specialists today look back at the early days of antibiotic resistance with a kind of nostalgic fondness. How much easier to deal with bacteria that produce a single enzyme that inactivates penicillin* than with a hospital full of patients infected with MRSA (for methicillin-resistant *S. aureus*), which doesn't just laugh at penicillin, but cephalosporin, ampicillin, and every other beta-lactam antibiotic? Or XDR TB (extensively drug-resistant tuberculosis), a bacillus that is unaffected by either isoniazid or rifampin, the more recent agents called fluoroquinolones, and at least one of these second-line drugs: capreomycin, kanamycin, or amikacin.

The human microbiome—the microorganisms in a particular environment—is largely composed of harmless or beneficial microbes, but it is also a perfect reservoir for spreading the genes for every imaginable form of antibiotic resistance.

There are any number of reasons for the explosive growth in the number and virulence of antibiotic-resistant bacteria—as late as the 1990s, fewer than 15 percent of hospital-acquired infections were

---

* Just to remind pharmacologists that bacteria are *always* ahead of the game: In 1972, dried soil was extracted from an exhibit that had lain untouched in the British Museum since 1689. Bacterial spores in the soil, which predated Alexander Fleming by some 250 years, nonetheless contained penicillinase, the enzyme responsible for penicillin resistance.

resistant to antibiotic treatment; acute-care hospitals today routinely report rates of 60 percent or more. The widespread use of subtherapeutic antibiotic treatment on livestock is unquestionably one of them. Overuse of antibiotics in animal feed leads to the creation and spread of antibiotic-resistant bacteria in poultry and meat consumed by the public.

Another driver of resistance is directly attributable to seven decades of overenthusiastic writing of prescriptions. Despite attempts to convince physicians to restrict the overuse of antibiotics from 1946 forward, the fact that antibiotics are so safe for any individual patient has persuaded hundreds of thousands of doctors to prescribe them for conditions for which they are almost always useless. A 1956 survey of doctors in North Carolina found that two-thirds of them, when presented with acute bronchitis—almost certainly a viral disease—prescribed antibiotics "indiscriminately to all patients. . . ." More than fifty years later, in 2010, 70 percent of emergency room doctors and 80 percent of primary care doctors were still prescribing antibiotic treatment for the disease. Patients bear some responsibility, too. Antibiotic prescriptions typically call for ten days of treatment in order to improve the odds that *all* the pathogenic bacteria causing the infection have been killed; since many antibiotics work by disrupting bacterial cell walls during cell division, and not all bacteria are dividing at the same time, it's critical to maintain a concentration of antibiotics until the entire pathogenic population has been exposed to them. In industrialized countries, though, up to 40 percent of patients fail to comply with the instructions they're given along with antibiotics; feeling better after a few days, they stop taking their medicine, thereby sparing the strongest and most resistant bacteria.

Part of the solution to antibiotic resistance is behavioral change. The Centers for Disease Control and Prevention have established "stewardship" programs intended to promote more judicious use of antibiotics, especially in hospitals, where 20 to 50 percent of prescriptions remain either inappropriate or unnecessary. Better and faster diagnostic tests can make it easier to distinguish the bacterial diseases that require antibiotics from the viral infections that don't.

Just as clearly, though, is the need for improving the incentives for developing new antibiotics, which is the logic behind GAIN. They've been a long time coming. In the thirty years after Proloprim appeared in 1969, not a single new class of antibiotics was licensed; every weapon against infectious disease was a derivative of an earlier one. And even since 2000, only two new classes have been approved for treatment: the oxazolidinones like Linezolid, which works by disrupting bacterial RNA translation, and Daptomycin, a cyclic lipopeptide that turns bacterial walls into Swiss cheese by literally changing their geometry.

It's not only that pharmaceutical companies aren't discovering new antibiotics. The old ones are disappearing, too. From 1938 to 2013, only 155 antibacterial compounds received FDA approval. Because of resistance, toxicity, and replacement by a newer-generation derivative, only 96 antibiotics remain available today. The decline isn't helped by the eagerness with which pharmaceutical firms are exiting the field. In 1988, thirty-two independent companies were actively researching antibiotics; during the 1990s, the number of companies that had received FDA approval for an antibiotic declined almost every year. It now stands at eleven, the lowest number since 1961. Some of this is the result of mergers, but not all. Though twenty-eight companies remain of the thirty-two operating in 1988, seventeen have left the field of antibacterial development altogether. Antibiotics built virtually every modern pharmaceutical company, but are now barely a rounding error in the industry's balance sheet.

If the current trend lines in the battle against infectious disease—every day, more resistance; every year, fewer new antibiotics—continue unchanged, the future takes on a distinctly scary cast: a world in which every puncture wound, or skin rash, or cough carries the risk of death from an unkillable, unstoppable bacterial pathogen. It wouldn't be precisely like the one that greeted George Washington on the last day of his life. It would be worse. Victims of bacterial infections in a completely antibiotic-resistant world would know precisely what was killing them. And would be utterly impotent to do anything about it.

Fortunately, those trend lines aren't set in stone. The Harvard-wide Program on Antibiotic Resistance (HWPAR) is developing novel methods for fighting bacterial pathogens, ones that don't actually kill bacteria, or even halt their reproduction, but degrade the structures that make the bacterium dangerous. A compound that attacks the toxins produced as part of the bacterial infection is the anti-infective warfare equivalent of defusing the enemy's artillery shells, rather than bombing the cannon themselves. Other researchers are developing ways to attack the source of antibiotic resistance: inhibiting the formation of the enzymes that penicillin-resistant bacteria use to disrupt the beta-lactam ring, for example. Michael Fischbach, at the University of California, San Francisco health campus, has discovered more than three thousand molecules within the human microbiome—the trillions of microorganisms that peacefully coexist inside our own bodies—that show antibiotic potential. Even better: Instead of relying on patience to await microbial innovation, Dr. Fischbach wrote a software program that could teach itself to recognize the patterns of successful antibiotic production in hundreds of existing microbial gene clusters.

And then there's GAIN. The tweaks it incorporates into the economics of antibiotic development—reducing the time, and therefore the costs, of bringing a drug to market; extending the patent life of drugs—are already bearing fruit. During a single four-month period in 2014—an even better year than 2012, with forty-four new drugs accepted—the FDA approved three distinct antibiotics as "qualified infectious disease products" specifically for the treatment of acute skin infections caused by MRSA: Dalvance, from the Chicago-based Durata Therapeutics; Sivextro, from Cubist Pharmaceuticals; and Orbactiv, developed by the Medicines Company of Parsippany, New Jersey. None of the new drugs represent a revolutionary advance, but a new protocol for antibiotic resistance, one that attracts the attention of the pharmaceutical industry, is one of the most hopeful signs imaginable in the battle against infectious disease.

The story of antibiotics, and the more general fight against disease, has alternated between unbridled optimism and dark foreboding. Every triumphal discovery—from Paul Ehrlich's arsenicals to Gerhard

Domagk's sulfanilamides to penicillin, streptomycin, and the broad-spectrum antibiotics—has been followed, sometimes in a matter of months, by a reminder that the enemy in this particular war may lose individual battles, but that the war against it is essentially eternal. Back in 1962, Ernst Chain bemoaned the adaptability of his old enemy, the staph bacterium, which "had again emerged as a dangerous disease against which he had no effective chemotherapeutic weapon." Bacterial pathogens, it seemed to him, were so adaptable that humanity's struggle against them was inevitably a losing game. The only plausible response to Chain's depressing conclusion came from his friend, the MIT chemist John Sheehan—the first person to chemically synthesize penicillin—who reminded him that, while the war against infectious disease was almost certain to go on forever, humanity wasn't any more easily defeated than the pathogens. "How about an expression of faith," Sheehan asked, "in the adaptability of the chemist?"

ACKNOWLEDGMENTS

It seems to me that acknowledgments appear so much more frequently in works of nonfiction than of imagination because of the widely held belief—rightly or wrongly—that nonfiction tends to be the product of many contributors, not simply one. This is a fairly romantic notion about storytelling in general, but it carries great weight with me. The subjects covered in each of my previous books were a complete mystery to me when I began researching, editing, and writing about them. To the degree that they are slightly less mysterious to me now is mostly due to the generous help of predecessors and contemporaries who left such clear footprints (in the case of the former) or signposts (for the latter) along the way.

The bibliography for this book contains hundreds of primary and secondary references. All were valuable, but a few of my predecessors were utterly essential for the writing of *Miracle Cure*, and the sheer number of times they are cited surely underlines this. Thanks to Eric Lax, Peter Pringle, Robert Root-Bernstein, Peter Temin, Robert Bud, Thomas Maeder, Eric Kandel, and Gwyn MacFarlane. The obituaries that appear in the series of *Biographical Memoirs of Fellows of the Royal Society* are indispensable for anyone trying to recreate *any* of the great scientific innovations of the mid-twentieth century.

A more focused level of help was forthcoming from this book's earliest days. Before *Miracle Cure* was even a proposal, David Jacobus, the founder of Jacobus Pharmaceutical Company, a man I'm privileged to call my friend, inspired me to find the theme of this project, which eventually took form as the title of Chapter Five: "To See the Problem

Clearly." My friend John Rosen read this manuscript at various points in its development and never failed to provide useful advice. Patti McKenna's knowledge of Merck's corporate archives, and Jeff Brand's of Pfizer's, were gifts of immeasurable value. Thomas Frusciano, university archivist at Rutgers' special collections, is owed thanks as well (not least for reminding me, after researching previous books took me from Scotland to Istanbul, that an impressive amount of the history of the antibiotic revolution took place less than an hour from my home in Princeton, New Jersey).

Two people in particular were vital to *Miracle Cure* in later stages of its gestation. No one in the world has done more to illuminate the history of pharmaceutical development than Mike Kinch, of the Institute for Public Health at Washington University, St. Louis (his actual title, as I write this, is associate vice chancellor and director, Center for Research Innovation in Business, and professor of radiation oncology, School of Medicine), and he has been extraordinarily generous in providing counsel and access to his research ever since he was managing the Center for Molecular Discovery at Yale. Julian West of Princeton University read the manuscript front to back and saved me from a frightening number of embarrassing chemical errors, and I am forever in his debt. As I hope is obvious, any remaining errors of fact are mine alone.

At Viking, Melanie Tortoroli served, as always, as a model editor, combining a powerful sense of narrative rhythm with a diligent advocacy on behalf of both the book and its readers . . . that is to say, you. At hundreds of points, she persuaded me to unpack complicated material so as to make it clearer, and to expand it, where needed, so as to make it more persuasive. If at any time you found yourself wanting either less or more on subjects like X-ray crystallography or patent disputes, I assure you it was one of the places where I overruled her advice. Her assistant editor, Georgia Bodnar, was diligent and knowledgeable about both the book and the process of publishing it. The book's copyeditor, Jane Cavolina, and its designers, Alissa Theodor and Nayar Cho (interior and jacket design, respectively) reminded me, yet again, of one of the most important and yet frequently ignored assets of what has become known as the "traditional" publishing model. Dozens of other members of the editorial and publishing team at Viking, led by Andrea Schulz and Brian Tart, guided the process with the imprint's customary care and grace.

Eric Simonoff, of William Morris Endeavor, has been my literary agent ever since I started writing. He was my friend before and remains so today. When I read, in the acknowledgments for other books, that this agent or that one is the best thing going, I smile to myself in the sure knowledge that only Eric's clients are actually privileged to write a line like that. For counsel, advocacy, and support, he may have the occasional equal, but I doubt it. Working with him has marked the most productive and enjoyable aspect of my entire professional life, and any description of all he has done for me would gain in implausibility in direct proportion to its accuracy.

Shortly after I began researching what would become *Miracle Cure*, I was diagnosed with a rare and highly aggressive form of cancer. I have, in consequence, acquired a debt of acknowledgment that I haven't had to think about much in previous projects. Dr. David August, Dr. Rebecca Moss, Dr. Elizabeth Poplin, and RN Joyce Plaza literally kept me alive long enough to complete *Miracle Cure*. So, too, did the hundreds of different researchers and clinicians at Novartis, Astra Zeneca, Bristol-Myers Squibb, and Bayer. It is, perhaps, a little ironic that a book that documents the birth of the modern pharmaceutical firm should have been so dependent on the products of its maturity.

For similar reasons, the traditional acknowledgment to family members for enduring the strains imposed by the writing life seems totally inadequate. The contribution of my wife, Jeanine, and my children—Alex, Emma, and Quillan—was above and beyond. As always, their thoughts have informed *Miracle Cure*, and there are innumerable grace notes they have shared with me. Seeking out both the images contained inside, and permission to reproduce them, was a job Jeanine cheerfully and professionally accepted.

But, of course, that says nothing about their real place in my life. They are the reason that it has been my privilege to love, and be loved, by some of the best people I have ever known.

William Rosen, April 2016

NOTES

PROLOGUE: "Five and a Half Grams"

1. a bronchial infection he acquired: (Grizzard, 2002) The actual disease—San Joaquin Valley Fever—was a serious one, caused by the fungus known as *Coccidiodes immitis*. As a fungal disease, it wasn't treatable by the penicillin that saved Anne Miller's life. (Tager, 1976)

ONE: "All the Worse for the Fishes"

5. "who was used to bleeding the people": (Grizzard, 2002)
6. "Do you understand me?": (Houting, 2011)
6. "to strangle a dog": (Hebert, 2009)
6. One thing that *didn't* kill: (Ellis, 2005)
8. *The Principles and Practices of Medicine*: (Osler, 1923) The first edition of Osler's classic was published in 1892, the last, posthumous one, four years after his death.
8. cared for them in hospitals: (Miller, T. S., 1984)
9. ulcerated gums and uncontrollable salivation: (Janik, 2014)
10. "Take a purified yellow Wax": (An American Physician, 1827)
10. therapy known as "swinging": (Belofsky, 2013)
10. "as you would a noisy dog or cat": (Janik, 2014)
10. "Every physician of experience": (Rosenberg, 1977)
13. so-called Plague of Athens: (Nelson, 2014)
13. "infection of imperceptible particles": (Thagard, 1996)
14. tuberculosis was responsible: (Jones, D. S., 2012)
14. Swiss mathematician Daniel Bernoulli: (Nelson, 2014)
16. "society [that] was cut in two": (Tocqueville, 1987)
18. Lavoisier described how sugar: (Barnett, 2003)
19. "the riddle of alcoholic fermentation": (Barnett, 2003)

19. "I do not think," he wrote: (Robbins, 2001)
21. Joseph Andreas von Stifft opened: (Kandel, 2012)
22. linked examination of living patients: (Kandel, 2012)
24. earth's 5×10^{30} bacteria: (Bratbak, 1985)
25. "It is not surprising that microbes": (Margulis, 1995)
25. "neither the cause": (Fitzgerald, 1923)
27. "Pasteur began with": (Koch, 1987)
28. "the most successful researcher": (Gradmann, 2001)
29. *"Hatred to Prussia. Vengeance"*: (Robbins, 2001)
30. new therapeutic technique for tuberculosis: (Gradmann, 2001)
30. *The Private Science of Louis Pasteur*: (Geison, 1995)
31. 12 million dollars today: Calculating the value of things over time is a notoriously tricky business, with half a dozen different methods in regular use. The simplest one just compounds increases in the Consumer Price Index over time, and gives us the $12 million figure. However, the price of labor has increased much more rapidly, and using the "income value" of the same 2.5 million gold francs—in 1888, each worth .29 grams of gold, while the U.S. dollar was pegged at 1.5 grams—produces the figure of $115 million. An even larger number, roughly $600 million, is the result of calculating those gold francs as their share of the French economy in 1888.
31. a girl named Julie-Antoinette Poughon: (Geison, 1995) For those interested in debating the appropriate level of adulation for Pasteur, the December 21, 1995, and April 4, 1996, issues of *The New York Review of Books* contain an exchange between Gerard Geison and Max Perutz on *The Private Science of Louis Pasteur*.
32. "Turning now to the question": (Thagard, 1996)
33. killed between 45 and 50 percent: (Godlee, 1918)
33. "Applying [Pasteur's] principles": (Thagard, 1996)
34. "the importance of the fact": (Lister, 1967)
35. "the tissue cells of man": (Jacobs, 1924)
35. the three Englishmen to make the cut: (Lawrence, 2004)
37. "all the worse for the fishes": (Holmes, *The Writings*, 1899) The original appeared in the *American Journal of Clinical Medicine* 18, no. 2, 163.

TWO: "Patience, Skill, Luck, and Money"

39. a technique used by the Lapps: (Bud, *Penicillin*, 2007)
39. the Harben Lectures given: (Witkop, 1999)

40. no qualms about partnerships: (Hager, 2006)
41. "an absolutely characteristic behavior": (Crivellato, 2003)
41. "the man with blue . . . fingers": (Hager, 2006)
42. Ehrlich had written a paper: (Volansky, 2009)
47. "I must . . . be no longer": (Witkop, 1999)
48. recognized by the Nobel committee: (Witkop, 1999)
53. "a biologically effective substance": (Riethmiller, 1999)
53. "uniform direction of research": (Van den Belt, 1997)
53. "often hammered on the anvil": (Van den Belt, 1997)
53. "endless combination game": (Van den Belt, 1997)
53. The atoxyl experiments were among: (Bosch, 2008)
56. test versions on rabbits, monkeys: (Van den Belt, 1997)
56. Desiderius Erasmus called it: (Rolleston, 1934)
56. "A night with Venus": (Frith, 2012)
57. "communities of interest": (Van den Belt, 1997)
57. The motto of Ehrlich's lab: (Hager, 2006)
57. "The material and mental support": (Van den Belt, 1997)
58. tended to introduce impurities: (Lloyd, 2005)
58. a magic buckshot: (Witkop, 1999)
60. more than 29,000 wounded: (Keegan, 1998)
60. onetime medical student: (Hager, 2006)
62. Carl Duisberg, the managing director: (Hager, 2006)
64. the bacteriologist exposed them: (Hager, 2006)
64. "By 1930 it was the universal opinion": (Galdston, 1943)
65. cured strep infections in mice: (Colebrook, "Gerhard Domagk," 1964)
67. "I had rather that those I esteemed": (Holmes, *The Writings*, 1899)
68. attacking up to three hundred: (Maeder, 1994)
68. mortality from childbed fever: (Hager, 2006)
69. sold in France by Poulenc as Stovarsol: (Ravina, 2011) Stovarsol, like
    Fourneau's cocaine derivative, which he named Stovaine, were plays
    on his own name. Fourneau, in English, means "furnace."
69. "I only had seven new products": (Hager, 2006)
70. "the German chemists' patents": (Hager, 2006)
70. original documents remain sealed away: (Hager, 2006)
71. a Japanese version was named: (Hager, 2006)
71. "I remember the astonishment": (Thomas, 1983)
73. the title of a book he wrote in 1929: (Carpenter, 2010)
73. number of advertised compounds: (Carpenter, 2010)
73. "Persenico" promised to combat: (Carpenter, 2010)
74. had no fraudulent intent: (Maeder, 1994)

75. "VENEREAL DISEASE 'CURE' KILLS": (Carpenter, 2010)

75. "nationwide race with death": (*New York Times*, October 25, 1937)

76. the existing law only allowed: (Carpenter, 2010)

76. Massengill eventually pleaded: (Maeder, 1994)

76. extended the "misbranding" violation: (Temin, 1980)

76. from 27 percent to 8 percent: (Hager, 2006)

77. "too polite to the Swedes": (Hager, 2006)

78. a 90 percent cure rate: (Hager, 2006)

THREE: "Play with Microbes"

80. Dr. Leo Schutzmacher . . . "dying of it": (Shaw, 1947)

81. "Jurisprudence and International Law": (Colebrook, "Almroth Edward Wright: 1861–1947," 1948)

82. On Praed Street: (Heaman, 2004)

83. a quarter of a million lines of poetry: (Dunhill, 2000)

84. "The physician of the future": (Colebrook, "Almroth Edward Wright: 1861–1947," 1948)

85. "to exploit the uninfected tissues": (Colebrook, "Almroth Edward Wright: 1861–1947," 1948)

85. "by uniting with the micro-organisms": (*New York Times*, March 31, 1907)

85. only twelve hundred soldiers died of it: (Colebrook, "Almroth Edward Wright: 1861–1947," 1948)

87. in a now-classic paper written for the *Lancet*: (Fleming, "Some Notes on the Bacteriology," 1915)

87. Fifteen percent of battlefield wounds: (Lax, 2004)

88. "that dissolved or killed the microbes": (Root-Bernstein, *Discovering*, 1989)

88. Fleming later revealed: (Hughes, 1979) and (Root-Bernstein, *Discovering*, 1989)

90. "On the Antibacterial Action": (Fleming, "On the Antibacterial Action," 1929)

90. the effect was observed after only two: (Lax, 2004)

91. created images . . . on Petri dishes: (Root-Bernstein, *Discovering*, 1989)

91. "his enjoyment came from games of all kinds": (Macfarlane, *Alexander Fleming*, 1984)

91. Fleming's penchant for games: (Root-Bernstein, "How Scientists Really Think," 1987)

92. a new source for lysozyme: This remains a minority opinion, even now, but is persuasively made in (Root-Bernstein, *Discovering*, 1989), to which I owe a great deal.

92. "Penicillin was not the first": (Fleming, "Penicillin," 1945)

93. "a thick blanket of *Penicillium*": (Clark, 1985)

94. predictably enough, Gram-negative: (Weidel, 2009) It wasn't until the 1960s, however, that anyone figured out why: The cell walls of all bacteria are composed of amino acids and sugars in a sort of mesh, a macromolecule known originally as murein, later as peptidoglycan. Some bacteria, however, have, in addition, an outer membrane made up of sugars and fats, formally a lipopolysaccharide, which sloughs off the violet stain.

94. Gram-positive/Gram-negative distinction: (Woese, 1987)

94. "All Fleming had to do": (Clark, 1985)

95. "The trouble of making it": (Abraham, "Howard Walter Florey," 1971)

95. "we knew very little": (Maurois, 1959)

98. "a first-rate man": (Abraham, "Howard Walter Florey," 1971)

98. "a rather nasty product": (Lax, 2004)

99. "make the peaks higher": (Jonas, 1989)

99. "future leaders in science": (Jonas, 1989)

101. "not a physically normal woman": (Lax, 2004)

101. "experiment a day, including Sundays": (Abraham, "Howard Walter Florey," 1971)

102. Britain's first Standards Laboratory: (Lax, 2004) and (Douglas, 1935)

103. Royal Charter that authorized collaboration: (Rasmussen, "The Moral Economy," 2004) and (Swann, *Academic Scientists*, 1988)

103. in a pure form from egg whites: (Abraham, "Ernst Boris Chain," 1983)

104. a factory that produced pure elements: (Abraham, "Ernst Boris Chain," 1983)

104. "paid or unpaid employment": (Lax, 2004)

104. a stipend of £250 annually: (Abraham, "Ernst Boris Chain," 1983)

105. "inexhaustible flow of ideas": (Clark, 1985)

105. quote page and volume: (Clark, 1985)

107. "his race and foreign origin": (Clark, 1985)

107. "principal motivating principle": (Clark, 1985)

107. "One! I shall want ten!": (Abraham, "Ernst Boris Chain," 1983)

108. "caused a terrific upheaval": (Clark, 1985)

108. "periodic fear attacks": (Lax, 2004)

108. "the chemical aspect of Pathology": (Lax, 2004)

108. "balances, micro balances": (Lax, 2004)

109. "I was a third-rate scientist": (Moberg, 1991)

109. "was wholly my conception and design": (Lax, 2004)

109. "Steel-bearing balls": (Heatley, 1939)

110. "Something seemed to click": (Clark, 1985)

110. Florey was equally insistent: (Lax, 2004)

112. "impracticable to grow": (Clark, 1985)

112. "a most versatile, ingenious": (Macfarlane, *Howard Florey*, 1979)

112. "He could improvise": (Macfarlane, *Howard Florey*, 1979)

114. "I have struggled to keep": (Lax, 2004)

115. "The application of Florey": (Bud, *Penicillin*, 2007)

115. Chain, as a refugee: (Abraham, "Ernst Boris Chain," 1983)

115. came through with only £25: (Lax, 2004)

117. he later called "laughably simple": (Bud, *Penicillin*, 2007)

117. no better than .02 percent pure: (Abraham, "Ernst Boris Chain," 1983)

117. "preparation from certain bacteria": (Clark, 1985)

118. "beautiful working hypothesis": (Clark, 1985)

118. "one mouse got up and staggered": (Abraham, "Howard Walter Florey," 1971)

119. "root and core and brain": (Churchill, 2002) This speech—the "We shall fight on the beaches" speech—was given before the House of Commons by the prime minister on June 4, 1940.

FOUR: "The People's Department"

120. a variety of purification techniques: (Clark, 1985)

123. the Dunn's elevator shut down: (Lax, 2004)

125. "In recent years . . . inhibited *in vitro*": (Chain, 1940)

127. "What [penicillin's] chemical nature is": (Lax, 2004)

127. "with my old penicillin": (Lax, 2004)

127. "I shall be surprised": (Lax, 2004) and (Macfarlane, *Howard Florey*, 1979)

128. "very undesirable that the Swiss": (Clark, 1985)

128. "I sympathize with your position": (Clark, 1985)

129. but the contaminants: (Clark, 1985) The process, pioneered by E. P. Abraham, involved successively exposing the filtrate to different powders that were attracted to distinct pigments in the fluid, then washing it in aluminum oxide until a powder appeared.

130. "was oozing pus everywhere": (Lax, 2004) A thank you to Mr. Lax, whose lively narrative informs many parts of this chapter.
131. closing in on a trillion dollars: (Maddison, 2006) Angus Maddison's indispensable tables calculating centuries' worth of world economic performance use an economic standard known as a "Geary-Khamis dollar," which is essentially the purchasing power of the U.S. dollar at a given point in time; in this case, the year 1990.
131. U.S. steelmakers rolled out: (Committee on Statistics, 1981) and (Woytinsky, 1953)
131. "some American mold or yeast raiser": (Bud, *Penicillin*, 2007)
132. "I have come to the conclusion": (Clark, 1985)
133. "scientific methods in strengthening": (National Academies, 2007)
133. corrected Fleming's misidentification: (Bud, *Penicillin*, 2007)
134. "It is precisely the people's Department": www.nal.usda.gov/lincolns-agricultural-legacy
136. "surplus agricultural commodities": (Finlay, 1990)
137. By 1944, this would be adopted: (Clark, 1985)
138. by 1929, Pfizer was selling: (Bud, "Innovators," 2011)
138. aluminum rotary fermenter: (Neushul, 1993)
138. Peoria team had a demonstration vat: (Bud, *Penicillin*, 2007)
139. "a favourable therapeutic response": (Abraham, "Howard Walter Florey," 1971)
139. four-and-a-half-year-old John Cox: (Lax, 2004)
140. "One could not trust": (Clark, 1985)
144. "for research on proteins and viruses": (Ferry, 1998)
145. "I've just come back from visiting Chain": (Ferry, 1998)
145. "immersed in a gummy fluid": (Ferry, 1998)
149. "the production of the drug": (Macfarlane, *Howard Florey*, 1979) and (Lax, 2004)
149. "a bouquet, at least": (Macfarlane, *Howard Florey*, 1979)
149. "I have now quite good evidence": (Clark, 1985)

FIVE: "To See the Problem Clearly"

151. "adequate provision for research": (Roosevelt, 1941)
152. protocols for funding through its six divisions: (Afflitto, 2012)
152. U.S. government had directly sponsored: (Starr, 1984)
153. the meeting, at which Bush presided: (Richards, A., 1964)

153. Years later, in 1957, Bush was named: (Zachary, 1997)

153. "it was possible to produce the kilo of material": (Neushul, 1993)

153. "every possible means of combating infection": (Hobby, 1985)

153. "A new pharmaceutical industry was born": (Neushul, 1993)

154. "Inventions cannot, in nature, be a subject of property": (Forman, 1900)

154. In 1921, three-quarters of the children: (Berg, 1973)

155. royalties well in excess of $7.5 million: (Bud, "Upheaval in the Moral Economy," 2008)

155. "Are Patents on Medicinal Discoveries": (Bud, "Upheaval in the Moral Economy," 2008)

155. Ernst Chain had proposed patenting: (Abraham, "Howard Walter Florey," 1971)

155. he was lobbying Dr. J. W. Trevan: (Abraham, "Ernst Boris Chain," 1983)

155. "a whole tremendous virgin field": (Clark, 1985)

156. "persisted in his 'money grubbing'": (Macfarlane, *Howard Florey*, 1979)

156. "gutters overrunning with": (Lax, 2004).

156. "a new and useful method": (Coghill, 1944)

157. It appears neither in his: (Neushul, 1993)

157. The key Coghill-Moyer patent: (Neushul, 1993)

157. "WHY NOT GO MERCK": (Lax, 2004)

159. roughly $96 million today: There are at least half a dozen ways of calculating the present value of a particular figure. This number uses the GDP deflator from the United States Bureau of Labor Statistics, found at data.bls.gov/cgi-bin/cpicalc.pl.

159. The Trading with the Enemy Act: (Steen, 2014)

161. the Powers-Weightman-Rosengarten Company: (Gortler, 2000)

161. At its dedication on April 25, 1933: (American Chemical Society, 1933)

161. Abbott Laboratories in 1914: (Kogan, 1963)

161. When Merck approached Richards: (Richards, A. N., 1949)

161. "saw in them no signs of the horns or tail": (Richards, A., 1964)

162. "Max was born with": (Gortler, 2000)

162. "to do research worthy": (Merck, "The Chemical Industry," 1935)

162. synthesized both vitamin B_6 and pantothenic acid: (Gortler, 2000)

162. "We're going to isolate every vitamin": (Tishler, 1983)

163. neither of which would license them to Merck: (Tishler, 1983)

163. "other firms who have made": (Lax, 2004)

164. in a toffee-flavored sugar cone: (Stevenson, 2014)

165. Alexander Hollaender, a researcher: (Bud, *Penicillin*, 2007)

165. new variety of the mold, known as Q-176: (Bud, *Penicillin*, 2007)

166. emergency physicians decided against debriding: (Cope, 1943)
166. Merck's Rahway facility packaged: (Sheehan, 1982)
167. "7 HOURS TO LIVE": (University of Pennsylvania, "Penicillin," 2002)
167. "GIRL, 20, DEAD AFTER REFUSAL": (*Chicago Tribune*, 1943)
168. A woman in Oklahoma City: (University of Pennsylvania, "Personal Correspondence," 2002)
169. "Richards . . . recognized the simple truth": (Clarke, 1949)
170. The office had recruited: (Bud, "Upheaval in the Moral Economy," 2008)
170. processing forty-two thousand surface cultures a day: (Bud, "Upheaval in the Moral Economy," 2008)
170. the CMR approved fifty-four contracts: (Richards, A., 1964) and (Neushul, 1993)
170. designated "scrambled facilities": (Federal Trade Commission, 1958)
170. After the war, the newly enriched: (Federal Trade Commission, 1958)
171. By 1944, it was: (Younkin, *Making the Market*, 2008)
171. The only comparable events: (Younkin, "A Healthy Business," 2010)
171. "Is it worth it?": (Flavell-White, 2010)
172. "followed every tank": (Bud, *Penicillin*, 2007)
172. Eli Lilly converted: (Kahn, 1976)
173. the giant Imperial Chemical Industries: (Bud, *Penicillin*, 2007)
173. the most powerful computing machinery: (Dodson, 2002)
173. Medical Research Council questioned her bill: (Bud, *Penicillin*, 2007)
174. they therefore "immediately accepted": (Dodson, 2002)
175. Hodgkin later presented Chain: (Clark, 1985)
176. "She radiated love: for chemistry": (Dodson, 2002)
176. Merck alone invested nearly $800,000: (Neushul, 1993)
177. by 1945, Germany was able to produce: (Bud, *Penicillin*, 2007)
178. I. G. Farben was investing: (Hayes, P., 1987)
179. nearly 40 percent *of all* industrial investment: (Hayes, P., 1987)
179. up to $1.5 billion in current dollars: (Hayes, P., 1987) and (Roth, 2011). The revenue in Reichsmarks—between 162 million and 351 million—has been translated using an exchange rate of 2.5 to the dollar, which was current in 1939.
179. Another $50 million was earned annually: (Hayes, P., 1987)
180. In 1944, it was barely one-fortieth: (Bud, *Penicillin*, 2007)
180. would eventually train five hundred clinicians: (Bud, *Penicillin*, 2007)
180. "the goal [is] to make penicillin so cheaply": (Bud, *Penicillin*, 2007), quoting Albert E. Elder, *The History of Penicillin Production* (1970).
181. "To you, Ernst Chain": (Holmberg, 1946)

SIX: "Man of the Soil"

182. Mineral Springs had only thirty-four beds: (Greenwood, 2008)
182. more than fifteen billion people: (Comas, 2013)
182. Skeletons exhumed from Egyptian gravesites: (Smith, I., 2003)
184. *Treponema pallidum*, the bacterium: (Todar, 2008–2012)
184. Since it's older than domestication: (Wirth, 2008)
187. European-style sanatorium in Saranac Lake: (Rothman, S. M., 1995)
188. "appear like fairies to gladden the hearts": (Chalke, 1962)
189. "Why do people go on dying of it?": (Shaw, 1947)
189. Though the BCG vaccine: (Roy, 2014)
191. Waksman's first fellow, Jackson Foster: (Major, 1949)
191. "production, purification . . . clinical trials": (Pringle, 2012)
192. "I know the actinomycetes can do better": (Pringle, 2012)
192. In a 1983 reminiscence: (Tishler, 1983)
192. "a chemical substance": (Pringle, 2012)
192. "We isolated one hundred thousand": (Schatz, 1993)
192. "real discovery was not streptomycin": (Greenwood, 2008) 152, quoting Bernard Davis
193. Schatz had been earning fifty dollars: (Tillitt, 1944)
194. "I worked day and night": (Schatz, 1993)
194. "a search for an antibiotic agent": (Schatz, 1993)
196. "medicine as a cooperative science": (Mayo Clinic, 2001–2015)
196. top veterinary pathologist . . . William Feldman: (Greenwood, 2008)
197. Streptomycin was effective: (Pringle, 2012)
198. Merck overruled his scientific staff: (Silcox, 1946)
199. Randolph Major put him to work: (Folkers, 1990)
199. "with some chicken wire": (Folkers, 1990)
199. "extremely imaginative, able, wonderful scientist": (Tishler, 1983)
200. twenty-three had survived: (Waksman, *The Conquest of Tuberculosis*, 1964)
200. "not the reason that we started working on it": This quote appears in dozens of biographies of Florey, but the original quote is from (Florey, 1967)
201. Merck scientists were allowed: (Tishler, 1983)
201. "You might care to wait": (Pringle, 2012)
203. In 1946, the War Department published: (Merck, "Biological Warfare," 1946)
204. twenty-two honorary doctoral degrees: (Daniel, 2001), 199
204. "new composition of matter" (Waksman, *The Conquest of Tuberculosis*, 1964)

204. "we gave you all the credit": (American Chemical Society, 2005)
205. "they were my tools, my hands, if you please": (American Chemical Society, 2005)
206. "you had nothing to do": (Waller, 2004)
207. "selected twelve patients": (Bhatt, 2010)
208. "we shall see how many funerals": (Doll, 1994)
208. so-called alternate allocation studies: (Podolsky, 2015)
209. Austin Bradford Hill: (Le Fanu, 1999)
211. "since the opportunity would never come again": (Le Fanu, 1999)
212. so had thirty-two who had received the treatment: (Le Fanu, 1999)
212. in 1946 Lehmann had published an article: (Lehmann, 1946)
212. "the combination of PAS with streptomycin": (MRC, 1949)
213. an almost unbelievable 80 percent: (Le Fanu, 1999)

SEVEN: "Satans into Seraphs"

216. It took nearly three years of testing: (Duggar, 1948)
217. "tinged with enthusiasm": (Podolsky, 2015) This was from Max Finland to Benjamin Carey, Lederle's director of laboratories.
217. "the most versatile antibiotic": (Podolsky, 2015)
217. shipping samples of their gold maker to 142,000 doctors: (Silberman, 1960), quoted in (Podolsky, 2015)
218. "We got soil samples": (Rodengen, 1999)
218. "because it came from the earth": (Rodengen, 1999)
219. "If you want to lose your shirt": (Podolsky, 2015)
219. Aureomycin accounted for 26 percent: (McEvilla, 1955)
220. "outstanding achievements in the art of organic synthesis": (Nobel Prize Foundation, 2014)
220. complicated molecule known as Vitamin B_{12}: (Woodward, 1973)
220. "None of us thought he was really that great": (Tishler, 1983)
220. it was Woodward who demonstrated that the beta-lactam structure: (Abraham, "Ernst Boris Chain," 1983)
221. facts "both clear and misleading": (Blout, 2001)
222. "by thought alone, deduced the correct structure": (Blout, 2001)
223. the FDA approved it: (Daemmrich, 2004)
223. on March 23, 1950: (Maeder, 1994)
223. seventy third-year medical students: (Mahoney, 1958)
223. By 1952, three hundred Pfizer reps: (Chandler, 2005)
225. By the early 1950s, more money: (Podolsky, 2015)

226. Lederle planned to launch Achromycin: (Maeder, 1994)
227. stabilized thereafter for another decade: (Temin, 1980)
227. Achromycin, its version of tetracycline: (Temin, 1980)
228. "Doctors Get a Heap of Comfort": (Grinnell, 1914)
228. the journal was carrying more advertising pages: (Temin, 1980)
229. virtually every issue of *JAMA* arrived: (Podolsky, 2015)
232. The mechanism for this: (Gaskins, 2002)
232. "dug residues out of the Lederle dump": (Jukes, 1985)
232. "Wonder Drug . . . had been found": (Laurence, 1950)
233. "The average weight": (Jukes, 1985)
233. what he called "Project Piglet" . . . as Terralac: (Rodengen, 1999)
235. Only the 31 showed resistance: (North, 1945)
235. May 1940, milk cows: (Bud, *Penicillin*, 2007)
236. milked before the penicillin: (Bud, *Penicillin*, 2007)
236. dosages were very small: (Graham, 2007)
236. biofilms, matrices that hold: (Davies, 2009)

EIGHT: "The Little Stranger"

238. "the only House in the West" . . . and Wormseed: (Kahn, 1976)
239. "profit on blood which has been donated": (Madison, 1989)
239. to three-quarters of the entire American market: (Chandler, 2005)
240. were "unreasonably high": (Madison, 1989)
240. "novel compound having antibiotic properties": (Bunch, 1952)
240. "looks at present quite hopelessly complex": (Todd, 1956)
240. The drug was Chloromycetin: (McEvilla, 1955)
241. "ether, sweet spirits of nitre": (Hoeffle, 2000)
241. Parke was its first president: (Mahoney, 1958)
241. "difficult to believe that one is under the influence of any drug at all":
     (Byck, 1974)
241. "the different varieties of coca": (Parke, Davis & Company, 1894)
241. "involving four thousand miles": (Parke, Davis & Company, 1894)
242. Davis even hired Freud himself: (Maeder, 1994) Freud's evaluation, for
     which he was paid an honorarium of 60 Dutch gulden—about $50—
     found that Merck's cocaine was, compared to Parke-Davis's, overpriced
     and difficult to find. (Van de Vijver, 2002)
242. "revive sexual existence": (Maine Academy of Medicine and Science,
     1897) Damiana is still marketed as an herbal aphrodisiac today.
242. first full-scale pharmaceutical research laboratory: (Chandler, 2005)

243. an émigré Basque farmer: (Bud, *Penicillin*, 2007)
243. more than seven thousand samples: (Burkholder, 1946)
243. a new antibiotic-producing organism: (Maeder, 1994)
243. a colony of S. *venezuelae*: (Greenwood, 2008)
244. "a teaspoon of soil": (Maeder, 1994)
244. The chemists at Parke-Davis gave it a nickname: (Maeder, 1994) Slightly later, on April 18, 1947, a team at the University of Illinois, under plant pathologist David Gottlieb and chemist Herbert Carter, isolated the same antibiotic source from composted soil at the university's Agricultural Experiment Station at Urbana. Their work was funded by Abbott Laboratories, Eli Lilly, Upjohn, and Parke-Davis itself, which was hedging its own bet. When, at a joint meeting, the Parke-Davis representatives realized that the Illinois culture, which they had named 8-44, was identical to their own "Little Stranger," Gottlieb's group withdrew and Parke-Davis left the consortium.
246. typhus killed as many as one German in ten: (Byrne, 2008)
246. typhus epidemic in the new Soviet Union: (Raoult, 2004)
246. Dr. Eugene Payne, had arrived in Bolivia: (Maeder, 1994)
248. Following fermentation . . . be extracted: (Maeder, 1994)
248. "The Greatest Drug Since Penicillin": (Podolsky, 2015)
248. more than $55 million in annual sales: (Maeder, 1994)
249. "Aplastic Anemia Due to Chloramphenicol": (Loyd, 1952)
250. more than four million people: (Maeder, 1994)
250. Only 61 of the victims: (Marks, 2012)
250. Benzene, for example, which is known to attack: (Smith, M., 1996)
251. "evidence was reasonably convincing": (Marks, 2012)
251. "We can't go on certifying": (Maeder, 1994)
252. A 1954 survey by the American Medical Association: (Temin, 1980)
252. "regulate the professional activities of physicians": (Marks, 2012)
252. nine-tenths of all prescribed medicines: (Moskowitz, 1957)
253. "sprawled on the public curb": (Maeder, 1994)
253. The risk of contracting aplastic anemia: (Temin, 1980)
255. "the 'drummer' of the drug house": (Brody, 2007)
255. "for instruction in the use of medicines": (Rasmussen, "The Drug Industry," 2005)
255. They weren't anything as low-rent: (Rothman, S., 2004)
255. "Detailing is, in reality, sales promotion": (Jones, T., 1940)
256. "The well-informed 'detail-man'": (Peterson, 1959)
257. "A generation of physicians": (Bean, 1955)
257. "The Professional Service Pharmacist's job": (Peterson, 1959)

257. the equivalent of a seminar in pharmacology: (Brody, 2007)
258. "If we put horse manure": (Maeder, 1994)
258. "Chloromycetin has been officially cleared": (Maeder, 1994)
258. "undoubtedly the highest compliment" . . . jobs worth having at Parke-Davis: (Maeder, 1994)
260. Gray babies frequently showed: (Maeder, 1994)
261. Thomas Hicks won the 1904 Olympic marathon: (Abbott, 2012)
262. "such conditions as . . . iron deficiency anemia": (Council on Drugs, 1960)
262. a comparable percentage of patients who took penicillin: (Temin, 1980)

NINE: "Disturbing Proportions"

265. "There is a need for writers": (Sexton, 2007)
265. "Prescription of antibiotics": (Lear, "Taking the Miracle Out," 1959)
267. "part of the blame": (Holmes, *The Writings*, 1899)
267. "Established ethical drug companies": (Lear, "Taking the Miracle Out," 1959)
267. "Sigmamycin . . . the antibiotic therapy of choice": (Podolsky, 2015)
267. "the most factual": (*Saturday Review*, 1959)
268. "attention to drug ads": (*Saturday Review*, 1959)
269. "Who better than the pharmaceutical industry": (Podolsky, 2015)
270. "Where my income comes from": (Lear, "The Certification of Antibiotics," 1959)
270. barely enough to pay his income tax: (Bud, *Penicillin*, 2007)
270. paid directly to Henry Welch: (Maeder, 1994)
270. "The February issue": (Podolsky, 2015)
272. "simultaneous and varied attack": (Podolsky, 2015)
272. sixty-one fixed-dose combination antibiotics: (Podolsky, 2015)
272. "with other useful therapeutic agents": (Podolsky, 2015)
273. randomized clinical trials made measuring: (Carpenter, 2010)
273. "The final verdict on the value": (Podolsky, 2015)
274. "advertising of soaps and tooth pastes": (Podolsky, 2015)
276. "none-too-happy picture": (Blair, 1938)
276. "sold twenty of [the] fifty-one": (Maeder, 1994)
276. margins were as high as 27 percent: (Bud, "Antibiotics, Big Business," 2005)
278. "he who buys does not order": (Congressional Quarterly, 1960)
278. "SENATE PANEL CITES MARK-UP ON DRUGS RANGING TO 7,079%": (Finney, 1959)

278. "about 10,000 percent": (Congressional Quarterly, 1960)
279. "brainwashing" tactics and "perverted marketing attitudes": (Congressional Quarterly, 1960)
279. Dirksen described Kefauver: (Silverman, 1974)
279. an antibiotic would be useless at any price: (Bud, *Penicillin*, 2007)
280. that seemed to underplay its risks: (Maeder, 1994)
280. "As a physician I blush with shame": (Podolsky, 2015)
282. "meaningless pseudoscientific jargon": (Kuehn, 2010)
282. Eight West German pediatric clinics: (Mintz, 1962)
283. distributed more than 2.5 million thalidomide pills: (McFadyen, 1976)
284. "'HEROINE' OF FDA": (Mintz, 1962)
284. "This is the story of how": (Mintz, 1962)
284. "a rare combination of factors": (Stephens, 2001)
286. A Kansas City woman: (Swann, "FDA and the Practice of Pharmacy," 1994)
287. "clear-cut method of distinguishing": (Reilly, 2006)
287. "on the basis of opinions generally held": (Reilly, 2006)
287. required the FDA to review every drug: (Maeder, 1994)
288. "adequate and well-controlled studies": (Junod, 2014)
288. "almost impossible to find": (Carpenter, 2010) The best-documented examples of this were Kelsey's denial of IND status—FDA permission to ship a drug across state lines to clinical investigators before it has been approved—to both dimethyl sulfoxide (DMSO), whose popularity persists in the world of alternative medicine, and LSD.

EPILOGUE: "The Adaptability of the Chemist"

298. GlaxoSmithKline, between 1995 and 2001: (Payne, 2007)
299. the FDA approved only three: (Porter, 2014)
299. more than $400 million: (DiMasi, 2003). Joseph DiMasi and his team at Tufts University added to the $400 million figure the cost of capital over the course of the drug's development; at a rate of 11 percent (which was itself fairly controversial), this doubled the cost, to about $800 million. With inflation, this is the billion-dollar price tag widely quoted today.
299. R & D cost of $43.4 million: (Light, 2011)
299. "there is no reasonable estimate available": (Light, 2011)
301. "It is not difficult to make microbes": (Fleming, "Penicillin," 1945)
301. Bacterial spores in the soil: (*New Scientist*, 1972)

302. "indiscriminately to all patients": (Bud, *Penicillin*, 2007)
302. 80 percent of primary care doctors: (Seppa, 2014)
302. up to 40 percent of patients fail to comply: (Kardas, 2006)
303. only 96 antibiotics remain available today: (Kinch, 2014)
303. It now stands at eleven: (Kinch, 2014)
304. Dr. Fischbach wrote a software program: (Zimmer, 2014)
305. "the adaptability of the chemist": (Bud, *Penicillin*, 2007)

BIBLIOGRAPHY

Abbott, K. "The 1904 Olympic Marathon May Have Been the Strangest Ever." *Smithsonian*, August 7, 2012.

Abraham, E. "Howard Walter Florey, Baron Florey of Adelaide and Marston, 1898–1968." *Biographical Memoirs of Fellows of the Royal Society* 17 (November 1971): 255–302.

Abraham, E. "Ernst Boris Chain: 19 June 1906–12 August 1979." *Biographical Memoirs of Fellows of the Royal Society* 29 (November 1983): 42–91.

Afflitto, E. *Penicillin, Venereal Disease, and the Relationship Between Science and the State in America, 1930–1950.* Ann Arbor, MI: University Microfilms/ProQuest, May 2012.

Allan, N. "We're Running Out of Antibiotics." *Atlantic*, March 2014, 34.

American Chemical Society. "Merck Laboratory Dedication." *Chemical Engineering News* 11, no. 9 (May 10, 1933): 137.

American Chemical Society. "Selman Waksman and Antibiotics. National Historic Chemical Landmarks." American Chemical Society (2005). Retrieved June 5, 2014. www.acs.org/content/acs/en/education/whatis chemistry/landmarks/selmanwaksman.html.

An American Physician. *The Eclectic and General Dispensatory.* Philadelphia: Towar and Hogan, 1827.

Barnett, J. A. "Beginnings of Microbiology and Biochemistry: The Contribution of Yeast Research." *Microbiology* 149, no. 3 (March 2003): 557–67.

Bean, W. B. "Vitamania, Polypharmacy, and Witchcraft." *Archives of Internal Medicine* 96, no. 2 (August 1, 1955): 137–41.

Belofsky, N. *Strange Medicine: A Shocking History of Medical Practice Through the Ages.* New York: Penguin, 2013.

Berg, A. *The Nutrition Factor: Its Role in National Development.* Washington, DC: Brookings Institution Press, 1973.

Bhatt, A. "Evolution of Clinical Research: A History Before and Beyond James Lind." *Perspectives in Clinical Research* 1, no. 1 (January–March 2010): 6–10.

Blair, J. C. *Seeds of Destruction: A Study in the Functional Weaknesses of Capitalism.* New York: Covici Friede, 1938.

Blaser, M. J. *Missing Microbes: How the Overuse of Antibiotics Is Fueling Our Modern Plagues.* New York: Picador, 2015.

Blout, E. *Robert Burns Woodward: 1917–1979. Biographical Memoirs,* Vol. 80. Washington, DC: National Academy Press, 2001.

Bosch, F., et al. "The Contributions of Paul Ehrlich to Pharmacology: A Tribute on the Occasion of the Centenary of His Nobel Prize." *Pharmacology,* 82, no. 3 (October 2008): 171–79.

Bowden, M. E. *Robert Burns Woodward and the Art of Organic Synthesis.* Philadelphia: Chemical Heritage Press, 1992.

Bratbak, G. "Bacterial Biovolume and Biomass Estimations." *Applied and Environmental Microbiology* 49, no. 6 (June 1985): 1488–93.

Brody, H. *Hooked: Ethics, the Medical Profession, and the Pharmaceutical Industry.* Lanham, MD: Rowman & Littlefield, 2007.

Bud, R. "Antibiotics, Big Business, and Consumers: The Context of Government Investigations into the Postwar American Drug Industry." *Technology and Culture* 46, no. 2 (April 2005): 329–49.

Bud, R. *Penicillin: Triumph and Tragedy.* Oxford: Oxford University Press, 2007.

Bud, R. "Upheaval in the Moral Economy of Science: Patenting, Teamwork, and the World War II Experience of Penicillin." *History and Technology: An International Journal* 24, no. 2 (March 7, 2008): 173–90.

Bud, R. "Innovators, Deep Fermentation, and Antibiotics." *Dynamis* 31, no. 2 (2011): 323–41.

Bunch, R. L. Erythromycin, its salts, and method of preparation. U. S. Patent No. US 2653899 A, filed April 14, 1952, and issued September 29, 1953.

Burkholder, P. R. "Studies on the Antibiotic Activity of Actinomycetes." *Journal of Bacteriology* 52, no. 4 (October 1946): 503–4.

Byck, R. *The Cocaine Papers of Sigmund Freud.* Edited and with an introduction by Robert Byck. New York: Stonehill, 1974.

Byrne, J. P. *Encyclopedia of Pestilence, Pandemics, and Plagues,* Vol. 1, A–M. Westport, CT: Greenwood Press, 2008.

Cantrill, S. "The Greatest Chemist of All Time?" *Nature Chemist* (January 7, 2011).

Carpenter, D. *Reputation and Power: Organizational Image and Pharmaceutical Regulation at the FDA.* Princeton, NJ: Princeton University Press, 2010.

Chain, E., et al. "Penicillin as a Chemotherapeutic Agent." *Lancet* 236, no. 6104 (August 24, 1940): 226–28.

Chalke, H. "The Impact of Tuberculosis on History, Literature, and Art." *Medical History* 6, no. 4 (October 1962): 301–18.

Chandler, A. D. *Shaping the Industrial Century: The Remarkable Story of the Evolution of the Modern Chemical and Pharmaceutical Industries.* Cambridge, MA: Harvard University Press, 2005.

Chicago Tribune. "Girl, 20, Dead After Refusal of Penicillin; Had 19 Transfusions for Blood Disease." *Chicago Tribune,* August 26, 1943.

Churchill, W. "We Shall Fight on the Beaches." Great Speeches of the Twentieth Century. *The Guardian,* April 20, 2002.

CIBA. *Man and His Future: A CIBA Foundation Volume.* Edited by Gordon Wolstenholme. Boston: Little, Brown, 1963.

Clardy, J. F. "The Natural History of Antibiotics." *Current Biology* 19, no. 11 (June 2009): 437–41.

Clark, R. *The Life of Ernst Chain: Penicillin and Beyond.* London: Bloomsbury, 1985.

Clarke, H. J. *The Chemistry of Penicillin.* Princeton, NJ: Princeton University Press, 1949.

Coghill, R., et al. Method for Increased Yields of Pencillin. U.S. Patent No. 2423873 A, June 17, 1944.

Colebrook, L. "Almroth Edward Wright: 1861–1947." *Obituary Notices of Fellows of the Royal Society* 6, no. 17 (November 1948): 297–314.

Colebrook, L. "Gerhard Domagk." *Biographical Memoirs of Fellows of the Royal Society* 10 (November 1964): 39.

Comas, I., et al. "Out-of-Africa Migration and Neolithic Co-Expansion of *Mycobacterium tuberculosis* with Modern Humans. *Nature: Genetics* 45, no. 10 (October 2013): 1176–82.

Committee on Statistics. *Steel Statistical Yearbook.* Brussels: International Iron & Steel Institute, 1981.

Congressional Quarterly. "Subcommittee Investigates Drug Prices." *CQ Almanac 1960,* 16th ed. Washington, DC: Congressional Quarterly, 1960, 11-743–49.

Cope, O. M. "The Treatment of the Surface Burns." *Annals of Surgery* 117, no. 6 (June 1943): 885–93.

Council on Drugs. "Blood Dyscrasias Associated with Chloramphenicol (Chloromycetin) Therapy." *Journal of the American Medical Association* 172, no. 18 (April 30, 1960): 2044–45.

Crivellato, E., et al. "Paul Ehrlich's Doctoral Thesis: A Milestone in the Study of Mast Cells." *British Journal of Hematology* 123 (2003): 19–21.

Daemmrich, A. *Pharmacopolitics: Drug Regulation in the United States and Germany.* Philadelphia: Chemical Heritage Foundation, 2004.

Daniel, T. M. *Pioneers in Medicine and Their Impact on Tuberculosis.* Rochester, NY: University of Rochester Press, 2001.

Dauer, C., et al. "Mortality from Influenza, 1957–1958 and 1959–1960." *American Review of Respiratory Diseases* 83 (1961): 15–28.

Davies, J. "Antibiotic Resistance and the Future of Antibiotics." In D. A. Relman, *Microbial Evolution and Co-Adaptation: A Tribute to the Life and Scientific Legacies of Joshua Lederberg.* Washington, DC: National Academies Press, 2009, 158–92.

de Costa, C., et al. "American Resurrection and the 1788 New York Doctors' Riot." *Lancet* 377, no. 9762 (January 2011): 22–28.

Dever, L. A. "Mechanisms of Bacterial Resistance to Antibiotics." *Archives of Internal Medicine* 151, no. 5 (May 1991): 886–95.

DiMasi, J. H. "The Price of Innovation: New Estimates of Drug Development Costs." *Journal of Health Economics* 22, no. 2 (March 2003): 151–85.

Dodson, G. "Dorothy Mary Crowfoot Hodgkin, O.M., 23 May 1910–29 July 1994." *Biographical Memoirs of the Fellows of the Royal Society* 48 (December 1, 2002): 179–219.

Doll, R. "Austin Bradford Hill, 8 July 1897–18 April 1991." *Biographical Memoirs of the Fellows of the Royal Society* 40 (November 1994): 128–40.

Douglas, S. "Georges Dreyer: 1873–1934." *Obituary Notices of Fellows of the Royal Society* 1, no. 4 (December 1935): 568–76.

Duggar, B. "Aureomycin: A Product of the Continuing Search for New Antibiotics." *Annals of the New York Academy of Sciences* 51 (November 1948): 177–81.

Dunhill, M. *The Plato of Praed Street: The Life and Times of Almroth Wright.* London: RSM Press, 2000.

Ellis, J. J. *His Excellency, George Washington.* New York: Vintage, 2005.

Federal Trade Commission. *Federal Trade Commission Economic Report on Antibiotics Manufacture.* Washington, DC: U.S. Government Printing Office, 1958.

Ferry, G. *Dorothy Hodgkin: A Life.* London: Granta Books, 1998.

Finland, M. J. "Occurrence of Serious Bacterial Infections Since Introduction of Antibacterial Agents." *Journal of the American Medical Association* 170, no. 18 (August 1959): 2188–97.

Finlay, M. R. "The Industrial Utilization of Farm Products and By-Products: The USDA Regional Research Laboratories." *Agricultural History* 64, no. 2 (Spring 1990): 41–52.

Finney, J. "Senate Panel Cites Mark-Ups on Drugs Ranging to 7,079%." *New York Times,* December 8, 1959, 1.

Fisher, M. W. "The Susceptibility of Staphylococci to Chloramphenicol." *Archives of Internal Medicine* 105, no. 3 (March 1960): 412–23.

Fisher, R. *The Genetical Theory of Natural Selection.* Oxford: Clarendon Press, 1930.

Fitzgerald, J. "Louis Pasteur: His Contribution to Anthrax Vaccination and the Evolution of a Principle of Active Immunization." *California State Journal of Medicine* 21, no. 3 (March 1923): 101–3.

Flavell-White, C. "Pfizer's Penicillin Pioneers." *TCE Today* (February 2010): 54–55.

Fleming, A. "Some Notes on the Bacteriology of Gas Gangrene." *Lancet* 186, no. 4799 (August 1915): 376–78.

Fleming, A. "On the Antibacterial Action of Cultures of a Penicillium, with Special Reference to Their Use in the Isolation of *B. influenzae. The British Journal of Experimental Pathology* 10 (May 1929): 226–36.

Fleming, A. "Penicillin." Nobel Lecture. December 11, 1945. NobelPrize .org. Retrieved November 1, 2014. www.nobelprize.org/nobel_prizes /medicine/laureates/1945/fleming-lecture.pdf.

Florey, H. Interview by H. de Berg. Tape recording. April 16, 1967. National Library, Canberra.

Folkers, K. A. Interview with Karl August Folkers. By L. Gortler. July 6, 1990. Philadelphia: Chemical Heritage Society.

Forman, S. *The Life and Writings of Thomas Jefferson, Including All of His Important Utterances on Public Questions, Compiled from State*

Papers and His Private Correspondence. Indianapolis: Bobbs-Merrill Company, 1900.

Frith, J. "Syphilis: Its Early History and Treatment until Penicillin and the Debate on Its Origins." *Journal of Military and Veteran's Health* 20, no. 4 (2012).

Galdston, I. *Behind the Sulfa Drugs: A Short History of Chemotherapy.* New York: Appleton-Century, 1943.

Gaskins, H., et al. "Antibiotics as Growth Promotants: Mode of Action." *Animal Biotechnology* 13, no. 1 (2002): 29–42.

Geison, G. L. *The Private Science of Louis Pasteur.* Princeton, NJ: Princeton University Press, 1995.

Godlee, S. R. *Lord Lister.* London: Macmillan, 1918.

Gortler, L. "Merck in America: The First 70 Years from Fine Chemicals to Pharmaceutical Giant." *Bulletin for the History of Chemistry* 25, no. 1 (2000): 1–9.

Gradmann, C. "Robert Koch and the Pressures of Scientific Research: Tuberculosis and Tuberculin." *Medical History* 45 (2001): 1–32.

Graham, J. P. "Growth Promoting Antibiotics in Food Animal Production: An Economic Analysis." *Public Health Reports* 122, no. 1 (January–February 2007): 79–87.

Grayson, M. B. "Henry Welch, FDA, and the Origins of ICAAC." *Microbe* 5, no. 9 (September 2010): 382.

Greene, J. A. "Reform, Regulation, and Pharmaceuticals—The Kefauver-Harris Amendments at 50." *New England Journal of Medicine* 167 (October 18, 2012): 1481–83.

Greenwood, D. *Antimicrobial Drugs: Chronicle of a Twentieth Century Medical Triumph.* New York: Oxford University Press, 2008.

Grinnell. "Doctors Get a Heap of Comfort from Grinnell Gloves." Advertisement. *Journal of the American Medical Association* 63 (October–November 1914): 31.

Grizzard, F. E. *George Washington: A Biographical Companion.* Santa Barbara, CA: ABC-CLIO, 2002.

Hager, T. *The Demon Under the Microscope: From Battlefield Hospitals to Nazi Labs, One Doctor's Heroic Search for the World's First Miracle Drug.* New York: Harmony Books, 2006.

Hayes, J. "Notes on Forensic Medicine: Smell." *The Graveyard Shift* (2008). Retrieved January 22, 2013. www.leelofland.com/wordpress/jonathan-hayes-notes-on-forensic-medicine-smell/.

Hayes, P. "Carl Bosch and Carl Krauch: Chemistry and the Political Economy of Germany 1025–1945." *The Journal of Economic History* 47, no. 2 (June 1987): 353–63.

Heaman, E. *St. Mary's: The History of a London Teaching Hospital.* London: Longmans, 2004.

Heatley, N. B. "A New Type of Microrespirometer." *Journal of Biochemistry* 33, no. 1 (January 1939): 53–67.

Hebert, M. "Who Killed George Washington?" *Medical Gumbo* (February 16, 2009). Retrieved August 17, 2014. open.salon.com/blog/michael_hebert/2009/02/16/what_killed_george_washington.

Hobby, G. *Penicillin: Meeting the Challenge.* New Haven, CT: Yale University Press, 1985.

Hoeffle, M. L. "The Early History of Parke-Davis and Company." *Bulletin of the History of Chemistry* 25, no. 1 (2000): 25–32.

Holmberg, A., et al. *"Les Prix Nobel en 1945."* Stockholm: The Nobel Foundation, 1946.

Holmes, O. W. "On the Contagiousness of Puerperal Fever." *New England Quarterly Journal of Medicine* 1 (1843): 503–30.

Holmes, O. W. *The Writings of Oliver Wendell Holmes,* Vol. 9. Boston: Houghton Mifflin, 1899.

Houting, B. A. "Did George Washington's Doctors Hasten His Death?" *Constitution Daily,* August 30, 2011.

Hughes, H. W. *Alexander Fleming and Penicillin.* London: Priory Press, 1979.

Jacobs, W. A. (1924). "Certain Aspects of the Chemotherapy of Protozoan and Bacterial Infections." *Medicine* 3, no. 2 (1924): 165–93.

Janik, E. *Marketplace of the Marvelous: The Strange Origins of Modern Medicine.* Boston: Beacon Press, 2014.

Johnson, S. *Where Good Ideas Come From: The Natural History of Innovation.* New York: Riverhead Books, 2010.

Jonas, G. *The Circuit Riders: Rockefeller Money and the Rise of Modern Science.* New York: W. W. Norton, 1989.

Jones, D. S. "The Burden of Disease and the Changing Task of Medicine." *New England Journal of Medicine* 366 (June 21, 2012): 2333–38.

Jones, T. *Detailing the Physician: Sales Promotion by Personal Contact with the Medical and Allied Professions.* New York: Romaine Pierson, 1940.

Jonsen, A. R. *The Birth of Bioethics.* Oxford: Oxford University Press, 1998.

Jordan, D. P. *Napoleon and the Revolution.* New York: Palgrave Macmillan, 2012.

Jukes, T. H. "Some Historical Notes on Chlortetracycline." *Review of Infectious Diseases* 7, no. 5 (September–October 1985): 702–7.

Junod, S. W. "FDA and Clinical Drug Trials: A Short History." *Overviews of FDA History* (September 7, 2014). Retrieved July 4, 2015. www.fda .gov/AboutFDA/WhatWeDo/History/Overviews/ucm304485.htm.

Kahn, E. J. *All in a Century: The First Hundred Years of the Eli Lilly Company.* Indianapolis: Eli Lilly, 1976.

Kandel, E. *The Age of Insight: The Quest to Understand the Unconscious in Art, Mind, and Brain, from Vienna 1900 to the Present.* New York: Random House, 2012.

Kardas, P. "Noncompliance in Current Antibiotic Practice." *Infectious Diseases in Clinical Practice* 14, no. 4 (July 2006): s11–s14.

Keegan, J. *The First World War.* New York: Knopf, 1998.

Kinch, M., et al. "An Analysis of FDA-Approved Drugs for Infectious Disease: Antibacterial Agents." *Drug Discovery Today* 19, no. 9 (September 2014): 1283–87.

Koch, R. *Essays of Robert Koch.* Edited by K. C. Carter. New York: Greenwood Press, 1987.

Kogan, H. *The Long White Line: The Story of Abbott Laboratories.* New York: Random House, 1963.

Kuehn, B. M. "Frances Kelsey Honored for FDA Legacy." *Journal of the American Medical Association* 304, no. 19 (November 17, 2010): 2109–12.

Laurence, W. L. "'Wonder Drug' Aureomycin Found to Spur Growth 50%." *New York Times*, April 10, 1950, 1.

Lawrence, C. "Joseph Lister." In *Oxford Dictionary of National Biography*, edited by H. C. G. Matthew et al. Oxford: Oxford University Press, 2004.

Lax, E. *The Mold in Dr. Florey's Coat.* New York: Henry Holt, 2004.

Le Fanu, J. *The Rise and Fall of Modern Medicine.* New York: Little, Brown, 1999.

Lear, J. "Taking the Miracle Out of the Miracle Drugs." *Saturday Review*, January 3, 1959, 35–41.

Lear, J. "The Certification of Antibiotics." *Saturday Review*, February 7, 1959, 43–48.

Lear, J. "SR Drug Reports and the United States Senate." *Saturday Review*, December 12, 1959, 49–50.

Lear, J. "Public Health at 7 1/2 Per Cent." *Saturday Review*, June 4, 1960, 37–41.

Lehmann, J. "Para-Aminosalicylic Acid in the Treatment of Tuberculosis." *Lancet* 247, no. 6384 (January 1946): 15–16.

Light, D., et al. "Demythologizing the High Costs of Pharmaceutical Research." *BioSocieties* 6 (February 2011): 34–50.

Lister, J. "Antiseptic Principle in the Practice of Surgery." *British Medical Journal* 2, no. 5543 (April 1967): 9–12.

Lloyd, N. C. "Salvarsan—The First Chemotherapeutic Compound." *Chemistry in New Zealand* 69, no. 1 (2005): 24–27.

Loyd, E. "Aplastic Anemia Due to Chloramphenicol." *Antibiotics and Chemotherapy* 2, no. 1 (January 1952): 1–4.

Macfarlane, G. *Howard Florey: The Making of a Great Scientist*. Oxford: Oxford University Press, 1979.

Macfarlane, G. *Alexander Fleming: The Man and the Myth*. Cambridge, MA: Harvard University Press, 1984.

Macfarlane, G. "Howard Florey." In *Oxford Dictionary of National Biography*, edited by H. C. G. Matthew et al. Oxford: Oxford University Press, 2004.

Maddison, A. *The World Economy*. Paris: Development Centre of the Organisation for Economic Co-operation and Development, 2006.

Madison, J. A. *Eli Lilly: A Life, 1885–1997*. Indianapolis: Indianapolis Historical Society, 1989.

Maeder, T. *Adverse Reactions*. New York: William Morrow, 1994.

Mahoney, T. *The Merchants of Life: An Account of the American Pharmaceutical Industry*. New York: Harper & Brothers, 1958.

Maine Academy of Medicine and Science. "Therapeutics of a Country Doctor." *Journal of Medicine and Science* 4, no. 1 (December 1897): 284.

Major, R. T. "Selman Waksman—Scientist, Teacher, and Benefactor of Mankind." Presentation Address for 1949 Research Award by the American Pharmaceutical Manufacturer's Association, in Honor of Dr. Selman A. Waksman. June 9, 1949. Hot Springs, VA.

Margulis, L., et al. *What Is Life*. New York: Simon & Schuster, 1995.

Marks, H. M. "Making Risks Visible: The Science and Politics of Adverse Drug Reactions. In *Ways of Regulating Drugs in the 19th and 20th*

Centuries, edited by J. P. Gaudilliere et al. Houndsmills, Basingstoke, UK: Palgrave Macmillan, 2012, 97–121.

Maurois, A. *The Life of Alexander Fleming: Discoverer of Penicillin.* London: Jonathan Cape, 1959.

Mayo Clinic. "Mayo Clinic Information: History" (2001–2015). Mayo Clinic. Retrieved March 17, 2015. www.mayoclinic.org/careerawareness /mi-history.html.

McEvilla, J. D. *Competition in the American Pharmaceutical Industry.* Pittsburgh: University of Pittsburgh, 1955.

McFadyen, R. "Thalidomide in America: A Brush with Tragedy. *Clio Medica* 11, no. 2 (July 1976): 79–93.

McKelvey, M. *Evolutionary Innovations: The Business of Biotechnology.* New York: Oxford University Press, 1996.

McLaren, A. *Impotence: A Cultural History.* Chicago: University of Chicago Press, 2007.

Merck, G. W. "The Chemical Industry and Medicine." *Industrial and Engineering Chemistry* 27, no. 7 (July 1, 1935): 739–41.

Merck, G. W. "Biological Warfare: Report to the Secretary of War by Mr. George W. Merck, Special Consultant for Biological Warfare" (January 3, 1946). Washington, DC: War Department Bureau of Public Relations, 1946.

Meynell, E. "Some Account of the British Military Hospitals of World War I at Etaples, in the Orbit of Sir Almroth Wright." *Journal of the Army Medical Corps* 142 (1996): 43–47.

Miller, G. H. "Abrupt Onset of the Little Ice Age Triggered by Volcanism and Sustained by Sea-Ice/Ocean Feedbacks." *Geophysical Research Letters* 39 (January 31, 2012).

Miller, T. S. "Byzantine Hospitals." Dumbarton Oaks Papers 38 (1984).

Mintz, M. "'Heroine' of FDA Keeps Bad Drug Off Market." *Washington Post*, July 15, 1962, 1.

Moberg, C. L. "Penicillin's Forgotten Man: Norman Heatley." *Science* 253, no. 5021 (August 19, 1991): 734–35.

Moskowitz, M. "Wonder Profits in Wonder Drugs." *The Nation*, April 27, 1957, 357–60.

MRC. "Treatment of Pulmonary Tuberculosis with PAS and Streptomycin: A Preliminary Report." *British Medical Journal* (December 31, 1949): 1521.

National Academies. "Articles of Organization of the National Research Council" (June 15, 2007). National Academies. Retrieved June 4, 2015. www.nationalacademies.org/nrc/na_070358.html.

Nelson, K. E. "Early History of Infectious Disease." In K. E. Nelson, *Infectious Disease Epidemiology*, 3rd ed. Burlington, MA: Jones and Bartlett, 2014, 1–21.

Neushul, P. "Science, Government, and the Mass Production of Penicillin." *Journal of the History of Medicine and Allied Sciences* 48, no. 4 (October 1993): 371–95.

New Scientist. "Some Bacteria Choose to Live in a Pool of Penicillin." *New Scientist*, June 8, 1972, 546.

New York Times. "The New Hope for Tuberculosis: Discovery of 'Opsonins' Promises to Revolutionize Medicine." *New York Times*, March 31, 1907, 1.

New York Times. "Near End of Chase for Deadly Elixir: Government Agents Hope to Recover Today the Last of 700 Bottles." *New York Times*, October 25, 1937.

Nobel Prize Foundation. "The Nobel Prize in Chemistry 1965" (2014). NobelPrize.org. Retrieved April 30, 2015. www.nobelprize.org/nobel _prizes/chemistry/laureates/1965/.

North, E., et al. "Observations on the Sensitivity of Staphylococci to Penicillin." *Medical Journal of Australia* 2 (1945): 44–46.

O'Neill, L. A. "Immunity's Early-Warning System." *Scientific American*, January 2005.

Osler, W. S. *The Principles and Practice of Medicine, Designed for the Use of Practitioners and Students of Medicine.* New York: Appleton, 1923.

Paget, S. *Pasteur and After Pasteur.* London: A&C Black, 1914.

Parke, Davis & Company. *Therapeutic Notes*, Vol. 1. Detroit: Parke, Davis & Company, 1894.

Payne, D. J. "Drugs for Bad Bugs: Confronting the Challenges of Antibacterial Discovery." *Nature Reviews—Drug Discovery* 6, no. 1 (January 2007): 29–42.

Peterson, A. F. *Pharmaceutical Selling, "Detailing," and Sales Training*, 2nd ed. Scarsdale, NY: Heathcote-Woodbridge, 1959.

Podolsky, S. H. *The Antibiotic Era: Reform, Resistance, and the Pursuit of a Rational Therapeutics.* Baltimore, MD: Johns Hopkins University Press, 2015.

Porter, E. "A Dearth of Innovation for Key Drugs." *New York Times*, July 23, 2014, B1.

Pringle, P. *Experiment Eleven: Dark Secrets Behind the Discovery of a Wonder Drug*. New York: Walker & Co., 2012.

Raoult, D., et al. "The History of Epidemic Typhus." *Infectious Disease Clinics in North America* 18, no. 1 (March 2004): 127–40.

Rasmussen, N. "The Moral Economy of the Drug Company-Medical Scientist Collaboration in Interwar America." *Social Studies of Science* 34 (2004): 161–85.

Rasmussen, N. "The Drug Industry and Clinical Research in Interwar America: Three Types of Physician Collaborator." *Bulletin of the History of Medicine* 79 (2005): 50–80.

Ravina, E. *The Evolution of Drug Discovery: From Traditional Medicines to Modern Drugs*. New York: John Wiley & Sons, 2011.

Reichl, R. "The F.D.A.'s Blatant Failure on Food." *New York Times*, July 30, 2014.

Reilly, G. W. *The FDA and Plan B: The Legislative History of the Durham-Humphrey Amendments and the Consideration of Social Harms in the Rx-OTC Switch* (May 12, 2006). LEDA at Harvard Law School. Retrieved June 28, 2015. dash.harvard.edu/bitstream/handle/1/8965550/Reilly06.html?sequence=2.

Richards, A. "Production of Penicillin in the United States (1941–1946)." *Nature* 201 (February 1964): 441–45.

Richards, A. N. "Remarks Concerning the Relations Between Universities, Industry and Government with Respect to Medical Research." The Clinical Problems of Advancing Years (March 16, 1949).

Riethmiller, S. "Ehrlich, Bertheim, and Atoxyl: The Origins of Modern Chemotherapy." *Bulletin of the History of Chemistry* 23 (1999): 28–33.

Robbins, L. *Louis Pasteur and the Hidden World of Microbes*. Oxford: Oxford University Press, 2001.

Rodengen, J. L. *The Legend of Pfizer*. Fort Lauderdale, FL: Write Stuff Syndicate, 1999.

Rolleston, J. "Venereal Disease in Literature." *British Journal of Venereal Disease* 10, no. 3 (1934): 147–74.

Roosevelt, F. D. Executive Order 8807 Establishing the Office of Scientific Research and Development (June 28, 1941). The American Presidency

Project. Retrieved December 14, 2014. www.presidency.ucsb.edu/ws/index.php?pid=16137#.

Root-Bernstein, R. S. "How Scientists Really Think." *Perspectives in Biology and Medicine* 32, no. 4 (Summer 1987): 473–89.

Root-Bernstein, R. S. *Discovering: Inventing and Solving Problems at the Frontiers of Scientific Knowledge.* Cambridge, MA: Harvard University Press, 1989.

Rosen, W. *The Most Powerful Idea in the World: A Story of Steam, Industry, and Invention.* New York: Random House, 2010.

Rosenberg, C. E. "The Therapeutic Revolution: Medicine, Meaning and Social Change in Nineteenth Century America." *Perspectives in Biology and Medicine* 20, no. 4 (Summer 1977): 485–506.

Roth, K. H. *Development and Production of Synthetic Gasoline.* Translated by Nicholas Levis. I. G. Farbenindustrie AG in World War II (2011). Retrieved March 8, 2015. www.wollheim-memorial.de/en/entwicklung_und_produktion_von_synthetischem_benzin.

Rothman, S. *The Pursuit of Perfection: The Promise and Perils of Medical Enhancement.* New York: Vintage Books, 2004.

Rothman, S. M. *Living in the Shadow of Death: Tuberculosis and the Social Experience of Illness in American History*, 1st ed. Baltimore, MD: Johns Hopkins University Press, 1995.

Roy, A., et al. "Effect of BCG Vaccination Against *Mycobacterium tuberculosis* Infection in Children: Systematic Review and Meta-analysis. *British Medical Journal* 349, no. g4643 (August 2014): 1–11.

Sarrett, L. H. *Max Tishler 1906–1989: A Biographical Memoir.* Washington, DC: National Academies Press, 1995.

Saturday Review. Letters to the Editor. *Saturday Review*, January 24, 1959, 21–23.

Schatz, A. "The True Story of the Discovery of Streptomycin." *Actinomycetes* 4, no. 2 (August 1993): 27–39.

Seppa, N. "Low-Tech Bacteria Battle." *Science News*, October 4, 2014, 22–26.

Sexton, P. *Legends of Literature: The Best Articles, Interviews, and Essays from the Archives of Writer's Digest Magazine.* New York: Writer's Digest Books, 2007.

Shaw, G. *The Doctor's Dilemma, Getting Married, & the Shewing-Up of Blanco Posnet.* London: Constable & Co., 1947.

Sheehan, J., et al. "The Fire That Made Penicillin Famous." *Yankee Magazine*, November 1982, 125–203.

Silberman, C. E. "Drugs: The Pace Is Getting Furious." *Forbes*, May 1960, 140.

Silcox, H. "Production of Penicillin." *Chemical Engineering News* 24, no. 20 (October 1946): 2762–64.

Silverman, M., et al. *Pills, Profits, and Politics*. Berkeley: University of California Press, 1974.

Smith, I. *Mycobacterium tuberculosis* Pathogenesis and Molecular Determinants of Virulence. *Clinical Microbiology Reviews* 16, no. 3 (July 2003): 463–96.

Smith, M. Overview of Benzene-Induced Aplastic Anaemia. *European Journal of Haemotology—Supplementum* 60 (1996): 107–10.

Starr, P. *The Social Transformation of American Medicine: The Rise of a Sovereign Profession and the Making of a Vast Industry*. New York: Basic Books, 1984.

Steen, K. *The American Synthetic Organic Chemicals Industry: War and Politics, 1910–1930*. Charlotte: University of North Carolina Press, 2014.

Stephens, T., et al. *Dark Remedy: The Impact of Thalidomide and Its Revival as a Vital Medicine*. New York: Basic Books, 2001.

Stevenson, W. "Charles Pfizer." In W. J. Hausman, *Immigrant Entrepreneurship: German American Business Biographies 1720 to the Present*, Vol. 2. Washington, DC: German Historical Institute, 2014.

Swann, J. P. *Academic Scientists and the Pharmaceutical Industry*. Baltimore, MD: Johns Hopkins University Press, 1988.

Swann, J. P. "FDA and the Practice of Pharmacy: Prescription Drug Regulation Before the Durham-Humphrey Amendment of 1951." *Pharmacy in History* 36, no. 2 (1994): 55–70.

Tager, M. "John F. Fulton, Coccidioidomycosis, and Penicillin." *Yale Journal of Biology and Medicine* 49 (June 1976): 391–98.

Temin, P. *Taking Your Medicine: Drug Regulation in the United States*. Cambridge, MA: Harvard University Press, 1980.

Thagard, P. "The Concept of Disease: Structure and Change." *Communication and Cognition* 29 (1996): 445–78.

Thomas, L. *The Youngest Science: Notes of a Medicine Watcher*. New York: Viking, 1983.

Tillitt, M. H. "Army-Navy Pay Tops Most Civilians' Unmarried Private's Income Equivalent to $3,600 Salary." *Barron's National Business and Financial Weekly*, April 24, 1944.

Tishler, M. Interview with Max Tishler. By L. G. Heitzman. November 14, 1983. Philadelphia: Chemical Heritage Society.

Tocqueville, A. de. *Reflections: The Revolutions of 1848*. New Brunswick, NJ: Transaction Publishers, 1987.

Todar, K. "*Mycobacterium tuberculosis* and Tuberculosis." Todar's Online Textbook of Bacteriology (2008–2012). Retrieved March 13, 2015. textbookofbacteriology.net/tuberculosis_3.html.

Todd, A., et al. *Perspectives in Organic Chemistry*. New York: Interscience Publishers, 1956.

University of Pennsylvania. "Penicillin and the American Public." Health, Medicine, and American Culture 1930–1960 (2002). Retrieved March 6, 2015. ccat.sas.upenn.edu/goldenage/state/pub/sl_pub1.htm.

University of Pennsylvania. "Personal Correspondence: Penicillin." Health, Medicine, and American Culture 1930–1960 (2002). Retrieved March 6, 2015. ccat.sas.upenn.edu/goldenage/state/pub/letters/sl_pub _letters_index.htm.

Van de Vijver, G., et al. *The Pre-Psychoanalytic Writings of Sigmund Freud*. London: Karnac Books, 2002.

Van den Belt, H. *Spirochaetes, Serology, and Salvarsan: Ludwig Fleck and the Construction of Medical Knowledge About Syphilis*. Wageningen, Netherlands: Grafisch bedrijf Ponsen & Looijen b.v., 1997.

Volansky, R. "Paul Ehrlich: The Man Behind the 'Magic Bullet.'" *HemOnc Today* (May 25, 2009).

Waksman, S. A. U. S. Streptomycin and the Process of Preparation, Patent No. 2,449,866, September 21, 1948.

Waksman, S. A. *The Conquest of Tuberculosis*. Berkeley: University of California Press, 1964.

Waksman, S. A. *The Antibiotic Era: A History of the Antibiotics and of Their Role in the Conquest of Infectious Diseases and in Other Fields of Human Endeavor*. Tokyo: Selman Foundation of Japan, 1975.

Waller, J. *Leaps in the Dark: The Making of Scientific Reputations*. Oxford: Oxford University Press, 2004.

Weidel, W., et al. "Bagshaped Macromolecules—A New Outlook on Bacterial Cell Walls." In F. Nord, *Advances in Enzymology and*

Related Areas of Molecular Biology, Vol. 26. New York: John Wiley & Sons, 2009, 193–223.

Wirth, T., et al. "Origin, Spread and Demography of the *Mycobacterium tuberculosis* Complex." *PLoS Pathogens* 4, no. 9 (September 2008): 1–10.

Witkop, B. "Paul Ehrlich and His Magic Bullets—Revisited." *Proceedings of the American Philosophical Society* 143, no. 4 (December 1999): 540–57.

Woese, C. R. "Bacterial Evolution." *Microbiological Reviews* 55, no. 2 (1987): 221–71.

Woodward, R. "The Total Synthesis of Vitamin B$_{12}$." *Pure and Applied Chemistry* 33, no. 1 (January 1973): 145–78.

Woytinsky, E., et al. *World Population and Production: Trends and Outlook.* New York: Twentieth Century Fund, 1953.

Xue, K. "Superbug: An Epidemic Begins." *Harvard Magazine,* May–June 2014, 40–49.

Younkin, P. A. *Making the Market: How the American Pharmaceutical Industry Transformed Itself During the 1940s.* Berkeley, CA: Institute for Research on Labor and Economics, March 2008.

Younkin, P. A. "A Healthy Business: The Evolution of the U.S. Market for Prescription Drugs." PhD diss., University of California, Berkeley, 2010.

Zachary, G. P. *Endless Frontier: Vannevar Bush, Engineer of the American Century.* New York: The Free Press, 1997.

Zimmer, C. "We May Be Our Own Best Medicine." *New York Times,* September 16, 2014, D7.

INDEX